PROGRAMMES IN
ORGANIC CHEMISTRY

5. REACTIONS OF CARBONYL COMPOUNDS

F. D. GUNSTONE
B.Sc., Ph.D., D.Sc., F.R.I.C.

*Reader in Chemistry, St. Salvator's College,
The University, St. Andrews, Fife*

THE ENGLISH UNIVERSITIES PRESS LIMITED
ST. PAUL'S HOUSE · WARWICK LANE · LONDON EC4

First printed 1969

SBN 340 05391 7

Printed for The English Universities Press Ltd. by
Willmer Brothers Limited, Birkenhead

PREFACE

Certain basic tenets about programmes were expounded in the prefa to each of the first three volumes in this series and will not be repeat here.

This programme was first written in 1966 and has been revised on tl basis of its use by students in St Salvator's College. It assumes previo knowledge of certain basic principles of organic chemistry such as nor enclature, the concept of resonance, and of terms such as nucleophil and electrophilic. It is particularly suitable for junior members universities and technical colleges and may be of some value to tl senior members of sixth forms and to teachers wishing to refresh the knowledge of this area of chemistry.

The term carbonyl is used loosely by organic chemists. This does n confuse those who understand, but it often presents difficulty to tl beginner. The carbonyl group is $>C=O$; carbonyl compounds som times include all compounds containing this group, but sometimes on aldehydes and ketones. This confusion is not easily resolved but wh its existence is explained the difficulty largely disappears.

This programme is concerned with the broad range of carbonyl including aldehydes and ketones on the one hand and the carboxyl acids and their derivatives on the other. It is common to emphasise tl distinction between these two types of carbonyl compounds: whil recognising their differences I prefer to emphasise their similariti which arise from the carbonyl function common to them all. M approach is briefly set out in Part One (frames 1–16) and elaborat in the subsequent Parts. It is suitable for those who have not studie these compounds before but will be more useful to those who, wi some knowledge of the subject, have not previously recognised the clo relationship between all the carbonyl compounds.

Students normally take six to seven hours to complete this pr gramme though times range between three and nine hours. Test fram appended to each part of the programme (frames 16, 74–76, an 142–144) can be used as pre-tests and post-tests. Those who obtai high pre-test scores will not need to work through the programme.

Some critics may complain about faults of programme-technique. admit to some of these and it may be that this is better described as Workbook rather than a Programme. Nevertheless it has prove beneficial to many students in its pre-published form and it is hope that it will now be equally useful to many more students.

I acknowledge helpful discussion and criticism from a number of colleagues, particularly R. K. Mackie. I thank Professor J. I. G. Cadogan for permitting me to make these teaching experiments, and for encouraging me to do so, and Professor W. G. Overend and the E.U.P. for the opportunity to present my programmes to other students. Above all, I am indebted to the many students who have worked through this programme at its various stages and have continually encouraged me in this venture.

F. D. GUNSTONE

CONTENTS

INSTRUCTIONS TO THE READER

The material presented to you in this programme is divided into short sections, called frames. Each frame usually contains explanatory material and one or more questions which you are meant to answer. Sometimes you are expected to insert a missing word or structure indicated by a line, thus ——, sometimes to choose between alternative answers. The correct answers are given after the next question.

Questions are separated from answers by a dotted line and complete frames are separated from each other by a continuous line. Answers should be masked until you have answered the relevant questions.

Write the answer on a separate sheet of paper. This will allow you to re-use the programme on a future occasion. Check each answer carefully and if it is correct continue with the next frame. If it is wrong try to understand why this is so and, if necessary, read through the explanatory material again. Make a note of the numbers of incorrect answers so that you can try them again later.

Work through the programme at your own pace. Do not try to complete it in one session but do not break off part way through a section.

Programmes are designed primarily for learning new material and not for revising but you can help yourself here either by marking the most important sections of the programme as you work through it or by making separate notes on what you read as you would with an ordinary text-book.

How to use the Programme

Cover the first page with a sheet of paper about the same size as the page, then pull the top of the paper down to the first line ruled across the page so that the first item of information, or "frame", is exposed.

Having read this frame carefully, decide on your answers to the questions, and write these down.

Then pull the covering paper down to the next line ruled across the page, so that item or "frame" 2 and the correct response to "frame" 1 are exposed. Check your response(s) to question 1; then continue with "frame" 2.

PART 1. INTRODUCTION

1 A carbonyl group exists when **oxygen** is linked to **carbon** by a double bond:

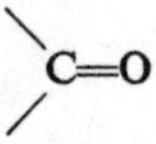

$$\diagdown C = O \qquad \text{carbonyl group}$$

This group occurs in several types of organic compounds which differ from one another in the nature of the two additional groups attached to the carbon atom.

In an **aldehyde** the carbonyl group is linked to hydrogen and to one other group which may be hydrogen, an alkyl group (aliphatic), or an aryl group (aromatic),

e.g.
$$\begin{array}{c} R \diagdown \\ \diagup C = O \\ H \end{array} \qquad \text{or R.CHO.}$$

A **ketone** has its carbonyl group attached to two other groups which can be either alkyl or aryl groups,

e.g.
$$\begin{array}{c} R \diagdown \\ \diagup C = O \\ R' \end{array} \qquad \text{or} \quad R.CO.R' \quad \text{or} \quad R_2C = O \text{ (when the two attached} \atop \text{groups are the same)}$$

Divide the following into four categories: aldehyde, ketone, other type of carbonyl compound, or not a carbonyl compound:

(i) $\begin{array}{c} H \diagdown \\ \diagup C = O \\ H \end{array}$
(ii) $\begin{array}{c} CH_3 \diagdown \\ \diagup C = O \\ C_6H_5 \end{array}$
(iii) $\begin{array}{c} C_2H_5 \diagdown \\ \diagup C = O \\ HO \end{array}$
(iv) $\begin{array}{c} C_2H_5 \diagdown \\ \diagup C = S \\ C_2H_5 \end{array}$

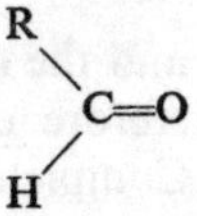

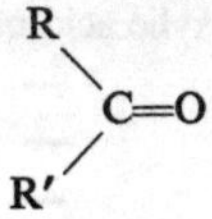

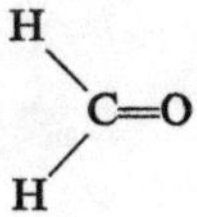

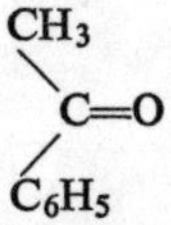

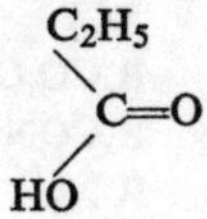

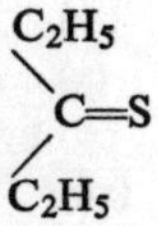

(v) $CH_3.CHO$ (vi) $CH_3.CH_2OH$ (vii) $Ph.COCl$ (viii) $Et.CO.Et$

(ix) $\begin{array}{c} O \\ \| \\ Et.C \\ \diagdown OEt \end{array}$
(x) [benzene ring with CHO]
(xi) $Et.CONH_2$
(xii) $CH_3.CH_2NO_2$

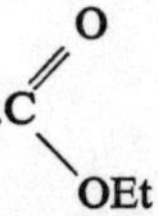

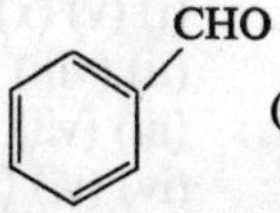

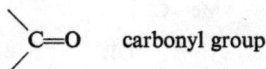

2 Some compounds containing a carbonyl group are neither aldehydes nor ketones. In most of these the carbonyl function is attached to *one* hydrogen atom, alkyl group, or aryl group, and to *one completely different group* such as OH, OR, NH_2, NHR, NR_2, Cl, and OCOR,

e.g.

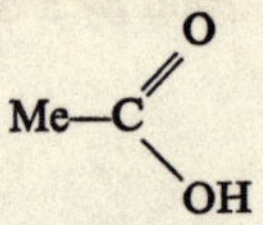

$$Me-C\overset{\displaystyle O}{\underset{\displaystyle OH}{\big|}}$$

carboxylic
acid

$$Me-C\overset{\displaystyle O}{\underset{\displaystyle OR}{\big|}}$$

ester

$$Me-C\overset{\displaystyle O}{\underset{\displaystyle NH_2}{\big|}}$$

amide

$$Me-C\overset{\displaystyle O}{\underset{\displaystyle Cl}{\big|}}$$

acid
chloride

$$Me-C\overset{\displaystyle O}{\underset{\displaystyle OCO.Me}{\big|}}$$

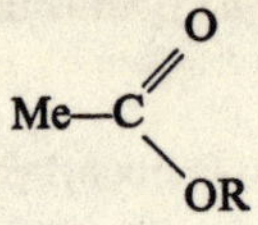

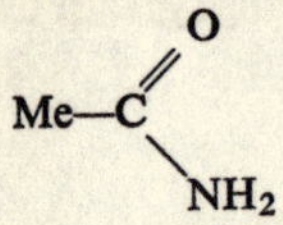

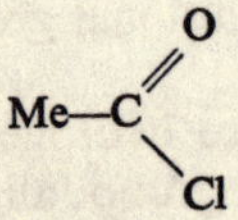

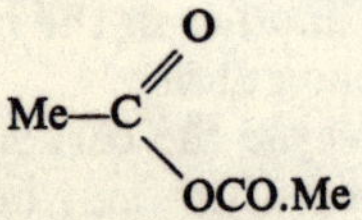

acid
anhydride

The first of these is a carboxylic acid and the others can be considered to be derived from it. They are therefore described collectively as **carboxylic acids and their derivatives.** Complete the following Table:

type of compound	general formula	classify as aldehyde, ketone, carboxylic acid or derivative, or no carbonyl group
e.g. primary alcohol	$R.CH_2OH$	no carbonyl group
(i) —	$R.CO_2H$	carboxylic acid or derivative
(ii) —	R.CHO	—
(iii) —	R_2CHOH	—
(iv) —	$R.CONH_2$	—
(v) —	R.COCl	—
(vi) —	RNH_2	—
(vii) —	R.CO.R	—
(viii) —	R.COOR'	—
(ix) —	RCl	—
(x) acid anhydride	R.CO.O.CO.R	—
(xi) hydrazide	$R.CONHNH_2$	—
(xii) —	ROR'	—

A1 aldehydes: (i) (v) (x)
    ketones: (ii) (viii)
    other carbonyl compounds: (iii) (vii) (ix) (xi)
    not a carbonyl compound: (iv) (vi) (xii)

3 The aldehydes and ketones on the one hand, and the carboxylic acids and their derivatives on the other, are sufficiently different to merit treatment in separate chapters of most text-books. It is unwise, however, to over-emphasise their differences for many of their reactions show common mechanistic features. These arise from the dominating influence of the —— group which they have in common.

Write a general structure of the type $\overset{A}{\underset{B}{\diagdown}}C{=}O$ for each of the following classes of compounds:

  (i) aldehydes, ketones.
  (ii) acids, esters, amides, acid chlorides, acid anhydrides.

A2 (i) carboxylic acid; (ii) aldehyde, aldehyde; (iii) secondary alcohol, no carbonyl group; (iv) amide, carboxylic acid or derivative; (v) acid chloride, carboxylic acid or derivative; (vi) amine, no carbonyl group; (vii) ketone, ketone; (viii) ester, carboxylic acid or derivative; (ix) alkyl chloride, no carbonyl group; (x) carboxylic acid or derivative; (xi) carboxylic acid or derivative; (xii) ether, no carbonyl group.

----

4 All these can be represented as $R{-}\overset{\overset{\textstyle O}{\|}}{C}{-}Z$

When Z is hydrogen we have an ——, when Z is an alkyl or aryl group a ——, and when Z is OH, OR′, NH_2, Cl, or O.CO.R we have a ——.

A3 carbonyl

$\overset{R}{\underset{H}{\diagdown}}C{=}O \qquad \overset{R}{\underset{R}{\diagdown}}C{=}O \qquad$ R=alkyl or aryl group

$\overset{R}{\underset{HO}{\diagdown}}C{=}O \qquad \overset{R}{\underset{R'O}{\diagdown}}C{=}O \qquad \overset{R}{\underset{H_2N}{\diagdown}}C{=}O \qquad \overset{R}{\underset{Cl}{\diagdown}}C{=}O \qquad \overset{R}{\underset{R.CO.O}{\diagdown}}C{=}O$

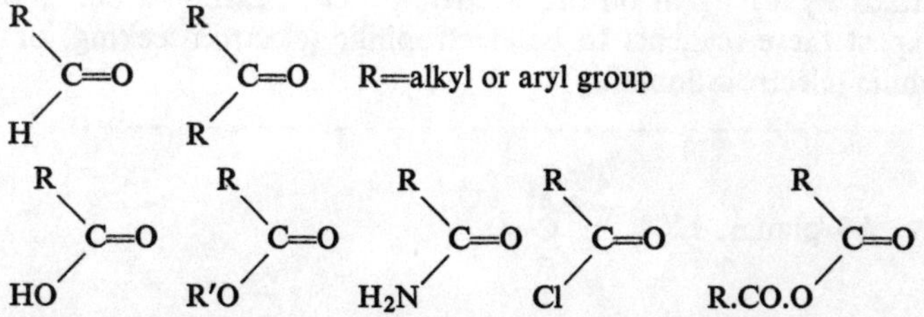

5 Since oxygen is strongly electron attracting, the C=O link is polarised in aldehydes and ketones. This can be represented in several ways:

$$R_2C{=}O \longleftrightarrow R_2\overset{+}{C}{-}\bar{O} \qquad R_2C\overset{\frown}{=}\ddot{O} \qquad R_2\overset{\delta+}{C}{=}\overset{\delta-}{\ddot{O}}$$

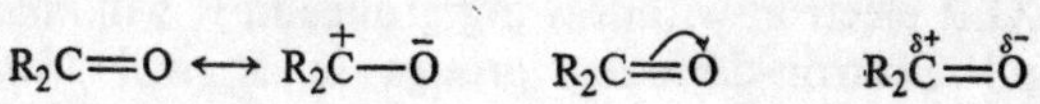

All of these representations serve to show that the carbonyl group has polar character with the oxygen atom electron-*rich/deficient* and the carbon atom electron-*rich/deficient*.

A4 aldehyde, ketone, carboxylic acid or derivative.

6 This can be expressed in a slightly different way. The carbonyl carbon atom is attached to *three* other atoms by σ bonds. These utilise sp^2 orbitals, lie in one plane, and are 120° apart. The remaining p

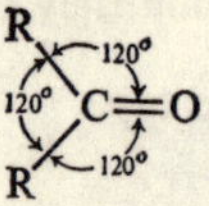

orbital of carbon overlaps with a p orbital of oxygen to form a π bond. The carbonyl double bond thus comprises one σ and one π bond. The oxygen atom is more strongly electron-attracting than the carbon atom and will therefore tend to pull the bonding electrons, and especially those in the more mobile π bond, closer to the oxygen. This uneven sharing of electrons is reflected in the dipole moment of aldehydes and ketones.

Formaldehyde is a *planar/non planar* molecule and the H—C—O angle is *109° 28′/120°*.

Insert appropriate charge signs on the δ symbols in

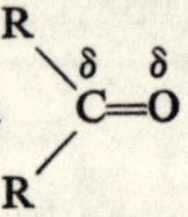

A5 The oxygen atom is electron-rich, the carbon atom is electron-deficient.

7 We shall discover that most reactions of a carbonyl group involve attack by a reagent on the electron-deficient carbon atom. Would you expect these reagents to be electrophilic (electron-seeking) or nucleophilic (electron-donating)?

A6 planar, 120°,

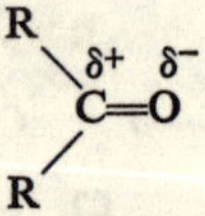

8 Reactions occur between the electron-donating nucleophilic reagent and the electron-deficient carbonyl carbon atom. Consequently in the ketone R.CO.R electron-withdrawing groups in R will *hinder/facilitate* reaction and electron-donating groups in R will *hinder/facilitate* reaction.

A7 Nucleophilic reagents react at electron-deficient centres.

9 This can also be stated the other way round. As most carbonyl reactions occur by interaction of a nucleophilic reagent with an electron-deficient carbon atom these reactions will be assisted by *electron-withdrawing/electron-donating groups* and hindered by *electron-withdrawing/electron-donating* groups in the carbonyl compound.

A8 Electron-withdrawing groups will increase the electron deficiency of the carbon atom and so facilitate the reaction. Electron-donating groups in R will decrease the electron deficiency of the carbon atom and so hinder the reaction.

---

10 In contrast to aldehydes and ketones, carboxylic acids and their derivatives necessarily contain an electron-donating group attached to the carbonyl group. Therefore they are less reactive towards those reagents which normally attack aldehydes and ketones. The electron-donating group Z affects the reactivity of the adjacent carbonyl group. The converse is also true. The carbonyl group affects the reactivity of the adjacent group Z and the properties of *alkyl* compounds ($R.CH_2Z$) differ from those of the corresponding *acyl* compounds (R.COZ). Complete the Table:

	alkyl derivatives		*acyl derivatives*	
	general structure	*general name*	*general structure*	*general name*
(i)	$R.CH_2Cl$	—	R.COCl	acyl or acid chloride
(ii)	$R.CH_2OH$	—	—	—
(iii)	$R.CH_2NH_2$	—	—	—
(iv)	$R.CH_2OR'$	—	—	—

The carbonyl group is *present/absent* in alkyl compounds and *present/absent* in acyl compounds. These two classes would therefore be expected to show *the same/different* reactivities.

A9 These reactions are assisted by electron-withdrawing groups and hindered by electron-donating groups in the carbonyl compound.

---

11 Most reactions of all types of carbonyl compounds can be fitted into the same general mechanistic sequence. Nucleophilic reagents (X^- or $\ddot{X}$) attack the carbonyl compound R.COZ at its electron-deficient carbon atom to give the species **(I)** which may react further in one of three ways. These are (i) addition of proton, (ii) elimination of X^- (this is equivalent to saying there is no reaction), and (iii) the elimination of Z^-.

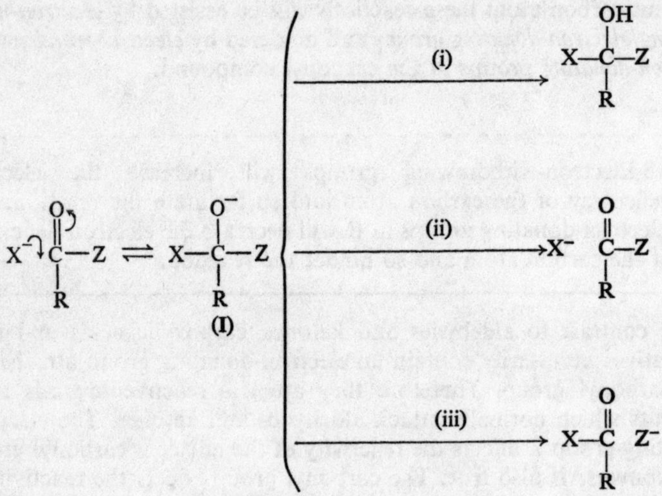

Sequence (i) involves addition of a —— to the ion (I) and the overall consequence of this reaction is **addition** of HX to the carbonyl compound. The addition reaction may be followed by a further reaction such as elimination of water but it is often possible to isolate the addition product.
Sequence (iii) involves elimination of —— from the ion (I) and the overall consequence of this reaction is **substitution** of the group Z by the group X. This substitution occurs at a (n) *saturated/unsaturated* carbon atom. Sometimes it is described as an addition-elimination reaction.

A10	(i)	alkyl chloride		
	(ii)	primary alcohol (or alkanol)	R.COOH	carboxylic acid
	(iii)	amine (or alkanamine)	$R.CONH_2$	amide
	(iv)	ether (or alkoxyalkane)	R.COOR′	ester

absent, present, different.

12 Aldehydes and ketones most commonly react by the **addition** sequence and we shall see that many of their reactions can be represented generally by the sequence:

$$\underset{R}{\overset{R}{>}}C{=}O \overset{X^-}{\rightleftharpoons} \text{——} \overset{H^+}{\rightleftharpoons} \text{——}$$

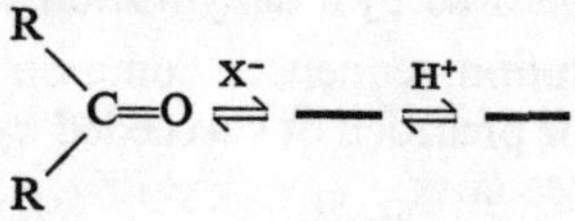

Carboxylic acid and their derivatives, on the other hand, most commonly react by the **substitution** sequence and their reactions are represented generally by:

$$R{-}\overset{\displaystyle O}{\underset{\displaystyle Z}{C}} \overset{X^-}{\rightleftharpoons} \text{——} \rightleftharpoons \text{——} + Z^-$$

- -

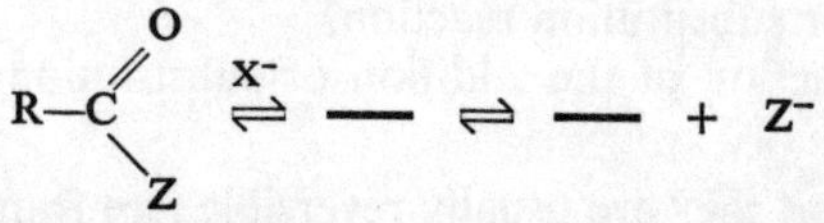

A11 proton, Z^-, unsaturated.

---

13 So we see that the two main classes of carbonyl compounds react in two-stage processes. The first step is common for both classes and involves attack by *nucleophilic/electrophilic* reagents at the *electron-deficient/electron-rich* carbon atom of the carbonyl group. With aldehydes and ketones this anion reacts with a proton in a(n) *addition/ substitution* reaction. With carboxylic acids and their derivatives the product usually results from elimination of —— and the net result is a(n) *addition/substitution* reaction.

- -

A12

$$\underset{R}{\overset{R}{>}}\overset{O^-}{\underset{X}{C}} \qquad\qquad \underset{R}{\overset{R}{>}}\overset{OH}{\underset{X}{C}}$$

$$R{-}\overset{\displaystyle X}{\underset{\displaystyle Z}{C}}{-}O^- \qquad\qquad R{-}\overset{\displaystyle X}{\underset{\displaystyle O}{C}}$$

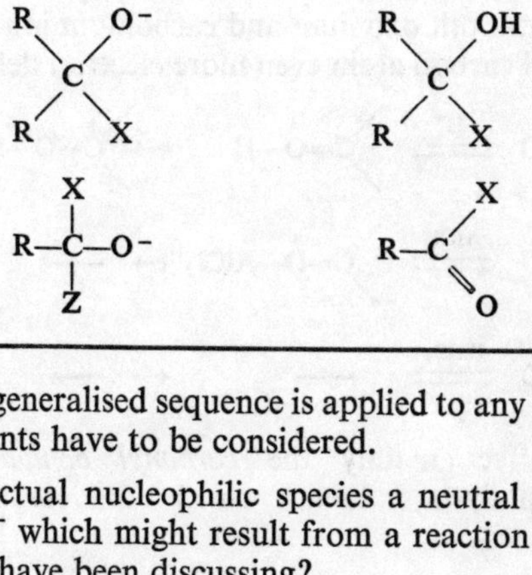

14 Before this generalised sequence is applied to any particular reaction two further points have to be considered.

(i) Is the actual nucleophilic species a neutral molecule $\ddot{X}$ or an anion X^- which might result from a reaction step preceding the ones we have been discussing?

(ii) The final product may not be any of those shown in the sequence discussed above (frame 11). These may only be intermediates which undergo further changes. This is particularly true of the addition reactions (commonly observed with —— and ——) which are often followed by a dehydration step.

The addition or substitution sequences common to most, if not all, carbonyl reactions *may* be preceded or succeeded by other reaction steps so the whole sequence becomes:

(i) formation of the actual nucleophile from the nucleophilic reagent (sometimes),
(ii) attack by nucleophile on the carbonyl group,
(iii) addition or substitution reaction,
(iv) further reaction of the addition or substitution product (sometimes).

Remembering that they are usually reversible (see frame 12) formulate steps (ii) and (iii) for

(a) reaction of CN^- with $CH_3.CHO$ (addition process);
(b) reaction of HO^- with $CH_3.CO_2Et$ (substitution process).

What steps other than (ii) and (iii) *may* be involved in the complete reaction sequence?

- -

A13 nucleophilic, electron-deficient, addition, Z^- (or an anion), substitution.

_____

15 Later we shall find that many of these reactions are catalysed by basic or acidic compounds. Basic catalysts usually increase the concentration and/or reactivity of the nucleophilic reagent. Acidic catalysts, both protonic and Lewis acids, react with carbonyl compounds to give reactive species with both oxonium and carbonium ion character. This makes the carbonyl carbon atom even more electron deficient.

e.g.

$$\ce{\chemfig{>C=O} ->[H^+] \chemfig{>C=\overset{+}{O}-H} <-> \chemfig{>\overset{+}{C}-O-H}}$$

$$\ce{\chemfig{>C=O} ->[AlCl3] \chemfig{>C=\overset{+}{O}-\overset{-}{A}lCl_3} <-> ——}$$

$$\ce{\chemfig{>C=O} ->[ZnCl2] —— <-> ——}$$

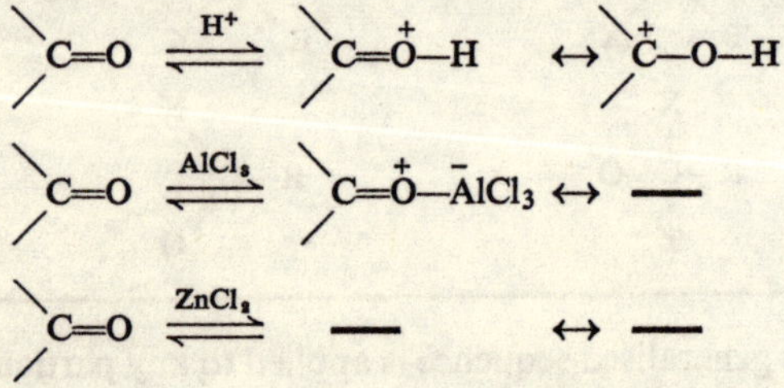

Basic catalysts affect mainly the *carbonyl compound/nucleophilic reagent;* acidic catalysts influence chiefly the *carbonyl compound/ nucleophilic reagent.*

- -

A14 aldehydes and ketones,

$$\underset{\substack{\| \\ O}}{CH_3.CH} + \bar{C}N \rightleftharpoons CH_3.\underset{\substack{| \\ O^-}}{CH}.CN \overset{H^+}{\rightleftharpoons} CH_3.CH(OH).CN$$

$$CH_3{-}\underset{\substack{| \\ OEt}}{\overset{\| \\ O}{C}} + \bar{O}H \rightleftharpoons CH_3{-}\underset{\substack{| \\ OEt}}{\overset{O^- \\ \curlywedge}{C}}{-}OH \rightleftharpoons CH_3.COOH + EtO^-$$

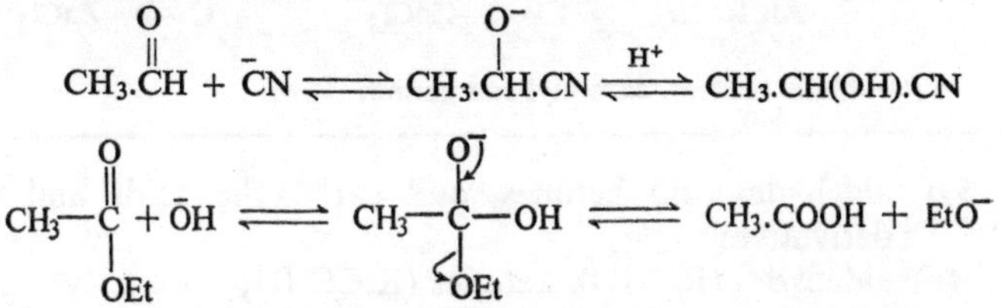

[If you get these right you have produced, by yourself, the mechanism of two important reactions: the formation of cyanhydrins and the alkaline hydrolysis of esters.]

Formation of the actual nucleophile from the nucleophilic reagent and further reaction of the addition or substitution product.

---

16 Test frame Answer all parts of this test before checking.
   (i) Compounds containing a C=O group are of two main types. What are these?
   (ii) Give the general name and formula of two groups of carbonyl compounds within each main type.
   (iii) The reactivity of carbonyl compounds is linked with the *electron-deficient/electron-rich* carbon atom which is subject to attack by *nucleophilic/electrophilic* reagents. Reactivity is enhanced by suitably placed electron-*withdrawing/donating* groups in the carbonyl compound.
   (iv) Alkyl (R.CH$_2$Z) and acyl (R.COZ) derivatives have *the same/ different* reactivity.
   (v) The group Z in an acyl compound *does/does not* affect the reactivity of the adjacent carbonyl group.
   (vi) The major reactions of carbonyl compounds are addition and substitution. Which type of reaction is particularly associated with which type of carbonyl compound?
   (vii) Formulate the anionic species formed by interaction of a carbonyl compound R.COZ and a nucleophilic reagent X$^-$ and then show how this anion undergoes addition or elimination.
   (viii) What additional steps may be involved in the whole reaction sequence?
   (ix) What sort of catalysts are commonly used?
   (x) Do these affect the nucleophilic reagent or the carbonyl compound or both?

A15

$$\underset{/}{\overset{\backslash}{C}}{-}O{-}\overset{-}{A}lCl_3 \qquad \underset{/}{\overset{\backslash}{C}}{=}\overset{+}{O}{-}\overset{-}{Z}nCl_2 \qquad \underset{/}{\overset{\backslash}{\overset{+}{C}}}{-}O{-}\overset{-}{Z}nCl_2$$

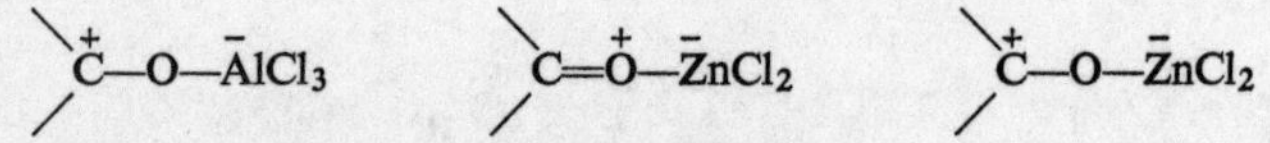

nucleophilic reagent, carbonyl compound.

A16 (i) aldehydes and ketones, and carboxylic acids and their derivatives

(ii) aldehydes (RCHO), ketones (R.CO.R);
two from: acids (R.COOH), esters (R.COOR′), amides (R.CONH$_2$), acid chlorides (R.COCl), acid anhydrides (R.CO.O.CO.R)

(iii) electron-deficient, nucleophilic, withdrawing

(iv) different

(v) does

(vi) aldehydes and ketones react mainly by addition, carboxylic acids and their derivatives mainly by substitution

(vii)

$$X{-}\underset{R}{\overset{O^-}{\underset{|}{\overset{|}{C}}}}{-}Z$$

$$\xrightarrow{\;+\,H^+\;} \quad X{-}\underset{R}{\overset{OH}{\underset{|}{\overset{|}{C}}}}{-}Z \quad \text{(addition)}$$

$$\longrightarrow \quad X{-}\underset{R}{\overset{O}{\underset{|}{\overset{\|}{C}}}}{+}\overset{-}{Z} \quad \text{(substitution)}$$

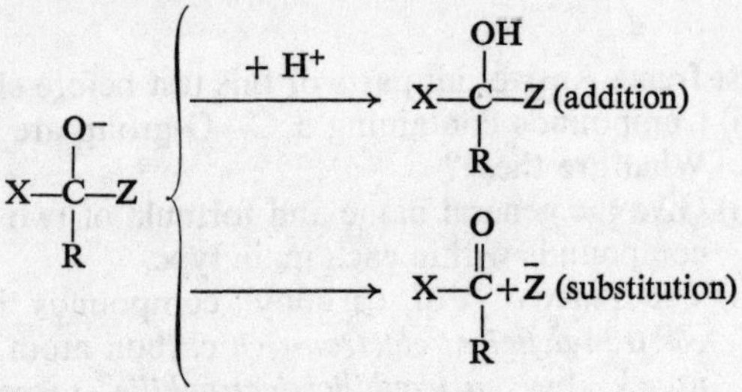

(viii) Production of the nucleophilic entity and further reaction of addition and substitution product

(ix) Acidic and basic catalysts

(x) Basic catalysts affect the nucleophilic reagent and acidic catalysts the carbonyl compound. (It will be shown later how acidic catalysts may affect the nucleophilic reagent also.)

PART 2 ALDEHYDES AND KETONES

17 Part 2 may take about three hours to complete and it is probably better not to attempt it in one session: break off at the end of any sub-section (see Contents).
The connection between these several sections may not be immediately obvious. All, however, are important reactions of aldehydes and ketones and follow the common mechanistic pathway of carbonyl addition.
Continue with frame 18.

---

18 (i) Addition of bisulphite
The addition of sodium bisulphite ($NaHSO_3$) to aldehydes and some ketones is a typical carbonyl addition reaction. The nucleophilic bisulphite ion attacks the carbonyl group and the intermediate anion undergoes proton transfer to form the product which is a sodium salt.

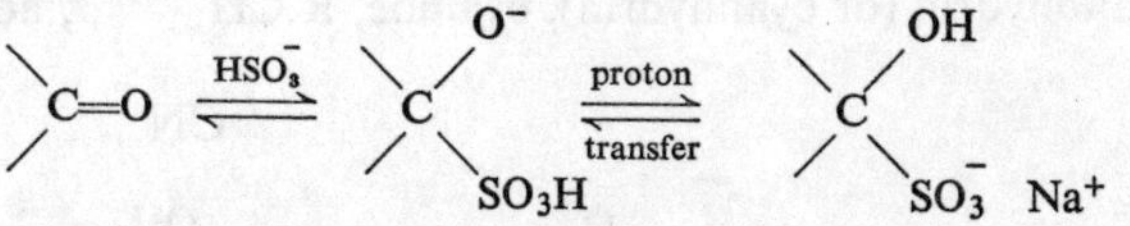

The proton gained by O^- does not come directly from the SO_3H group in the same molecule since proton transfer will occur *via* the aqueous solvent. Formulate the reaction between propionaldehyde and sodium bisulphite.

---

19 (ii) Addition of hydrogen cyanide
Aldehydes and ketones react with hydrogen cyanide to give *cyanohydrins*,

e.g.

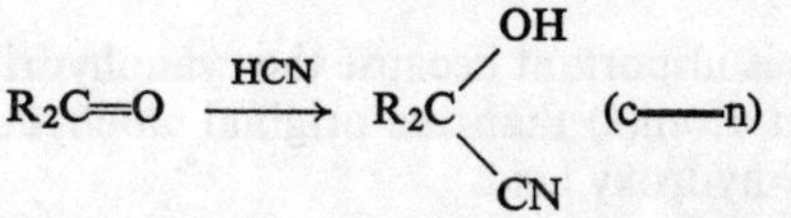

The reaction is effected usually by a solution of sodium or potassium cyanide containing a little mineral acid or with a solution of hydrogen cyanide containing a trace of alkali. Both these mixtures contain a buffered supply of CN^- and H^+ (the two ions needed in this reaction). Attack by the nucleophilic —— ion precedes the protonation stage thus:

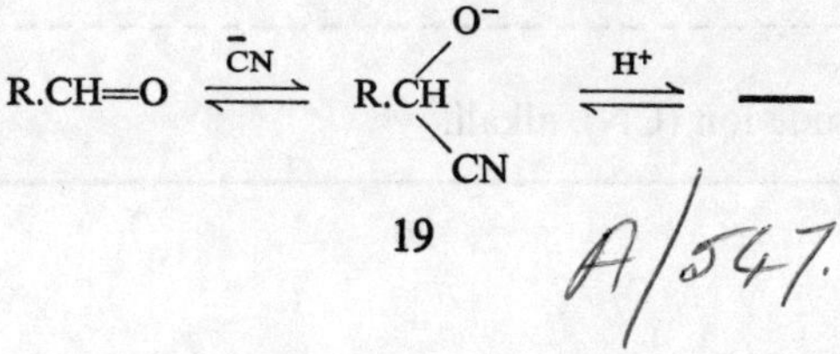

This is a(n) *addition/substitution* reaction of aldehydes and ketones. Formulate the reaction between acetone and (i) hydrogen cyanide and (ii) sodium bisulphite.

A18

$$Et.CH{=}O \xrightleftharpoons{HSO_3^-} Et.CH\begin{smallmatrix}O^-\\ \\SO_3H\end{smallmatrix} \rightleftharpoons Et.CH\begin{smallmatrix}OH\\ \\SO_3^-\ Na^+\end{smallmatrix}$$

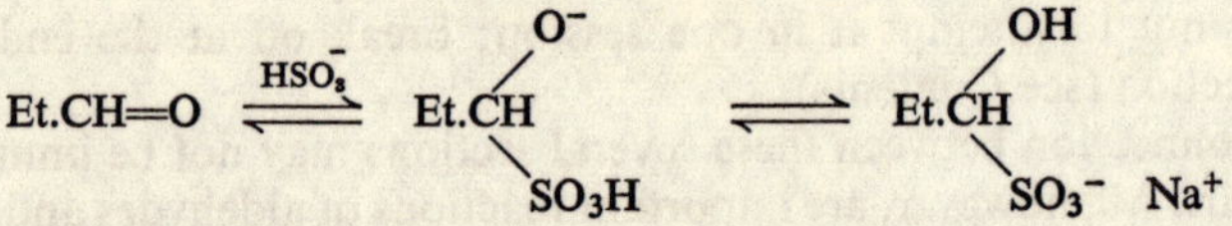

20 Reaction with hydrogen cyanide alone is very slow. The acid is not extensively dissociated and so the concentration of the nucleophilic reagent (*i.e.* ——) is very low. It is increased by addition of a little ——.

A19 cyanohydrin (or cyanhydrin), cyanide, $R.CH\begin{smallmatrix}OH\\ \\CN\end{smallmatrix}$, addition,

$$Me_2C{=}O \xrightleftharpoons{\bar{C}N} Me_2C\begin{smallmatrix}O^-\\ \\CN\end{smallmatrix} \xrightleftharpoons{H^+} Me_2C\begin{smallmatrix}OH\\ \\CN\end{smallmatrix}$$

$$Me_2C{=}O \xrightleftharpoons{HSO_3^-} Me_2C\begin{smallmatrix}O^-\\ \\SO_3H\end{smallmatrix} \rightleftharpoons Me_2C\begin{smallmatrix}OH\\ \\SO_3^-\ Na^+\end{smallmatrix}$$

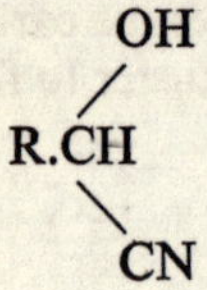

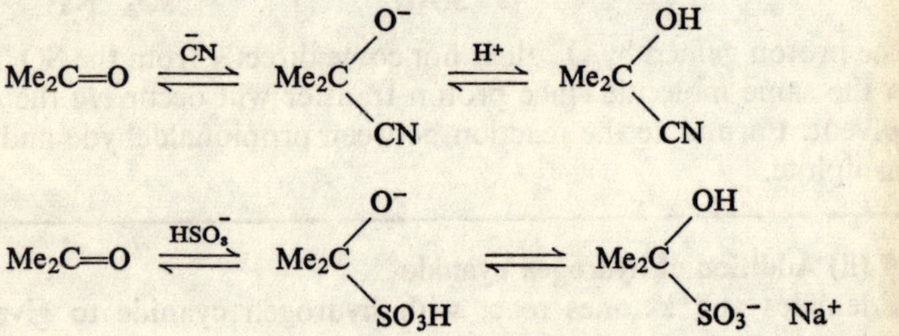

21 This reaction is important because the cyanohydrin contains *one/two more/less* carbon atom(s) than the original aldehyde and provides, by hydrolysis, an α-hydroxy acid.

$$R.CHO \xrightarrow{HCN} R.CH\begin{smallmatrix}OH\\ \\CN\end{smallmatrix} \xrightarrow{hydrolysis} \quad\text{——}$$

aldehyde   ——   carboxylic acid

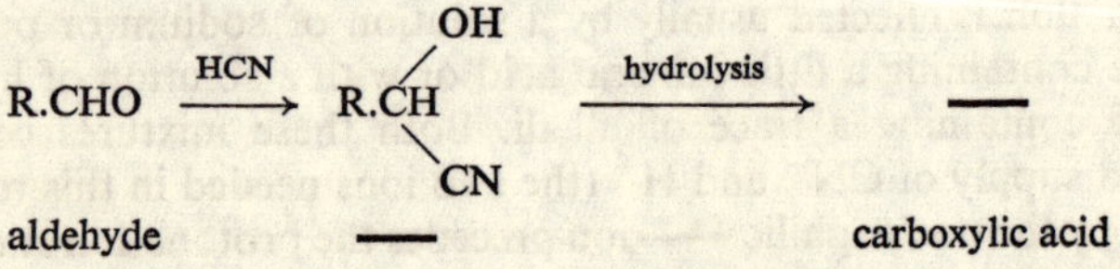

A20 cyanide ion ($\bar{C}N$), alkali.

22 (iii) Benzoin condensation

Aromatic aldehydes can react with cyanide ion in a different way yielding a ketol or hydroxy-ketone,

e.g.

$$2Ph.CHO \overset{\bar{C}N}{\rightleftharpoons} Ph.CO.CH(OH).Ph$$

benzaldehyde   benzoin

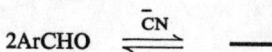

$$2ArCHO \overset{\bar{C}N}{\rightleftharpoons} \text{——}$$

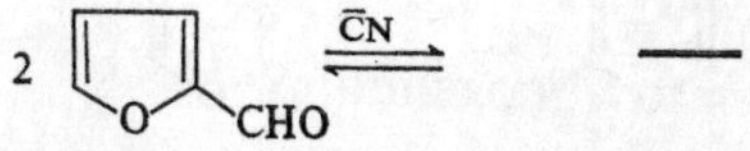

furfural   furoin

A21 one more, cyanohydrin, R.CH(OH).CO$_2$H

---

23 This reaction occurs in five stages:

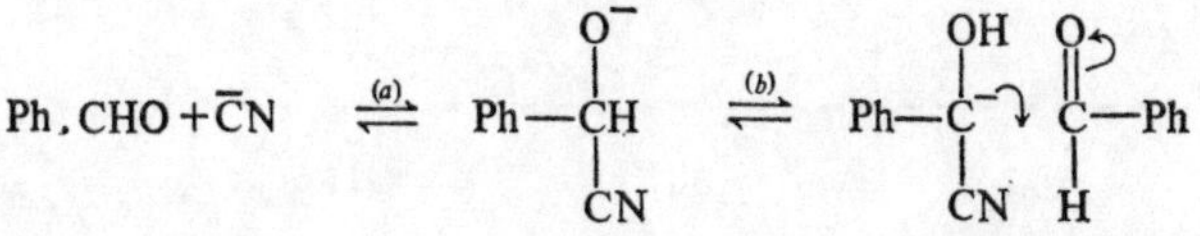

(II)   (III)

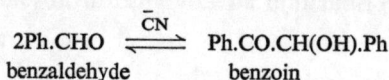

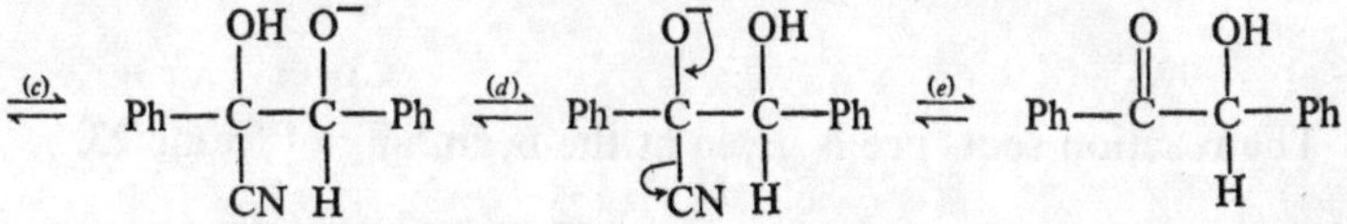

(IV)

Putting this in words we have:

(*a*) Reaction of aldehyde with —— ion to give the oxy-anion **(II)**.

(*b*) This ion is tautomeric and can exist as **(III)** following a proton shift from carbon to oxygen. Normally oxy-anions are much more stable than carbanions but **(III)** is stabilised by the adjacent aromatic ring and also by the cyano group, both of which permit delocalisation of the negative charge.

(*c*) The carbanion **(III)** is a nucleophilic reagent which now attacks a second aldehyde molecule in the usual way to give **(IV)**.

(*d*) The oxy-anion **(IV)** undergoes proton exchange *via* solvent.

(*e*) Elimination of —— ion yields the ketol (benzoin).

Notice that the cyanide ion used in the first stage of reaction is eliminated in the last stage and can therefore be described as a —— . This reaction involves two aldehyde molecules. Both are attacked by

nucleophilic reagents in the usual way: the first by —— ion and the second by a carbanion of structure ——.
Formulate in five stages the conversion of benzaldehyde into benzoin. (If you can't remember the sequence of events refer to the verbal description. Only consult the formulation already given if absolutely necessary.)

A22

Ar.CO.CH(OH).Ar 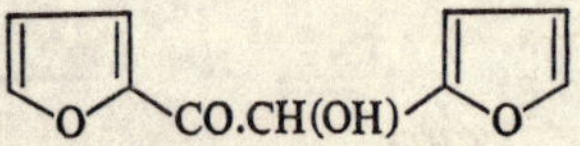

---

24 What are the structures of the reaction products from cyanide and each of the four aldehydes?

$CH_3.CHO$ p-$CH_3.C_6H_4.CHO$ 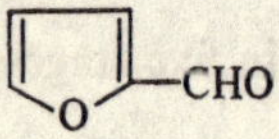 $CH_3.CH_2.CH_2.CHO$

A23 cyanide, cyanide, catalyst, cyanide, Ph—C⁻(OH)(CN)

$$\begin{array}{c} OH \\ | \\ Ph—C^- \\ | \\ CN \end{array}$$

The reaction sequence is given at the beginning of frame 23.

---

25 The formation of bisulphite addition compounds and of cyano-hydrins are the simplest addition reactions of aldehydes and ketones. Many other reactions of these compounds are essentially addition processes but some extra feature makes them a little more complicated and therefore more interesting.

(iv) Formation of acetals and ketals

Aldehydes (and to a lesser extent ketones) react with alcohols, usually in the presence of mineral acids or other acidic catalysts, to give first the unstable hemi-acetal and then the more stable acetal.

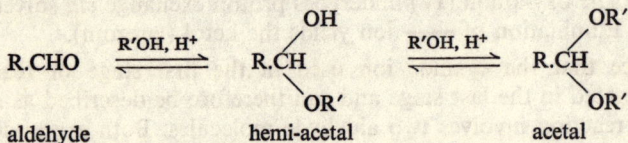

$$\underset{\text{aldehyde}}{R.CHO} \xrightleftharpoons{R'OH,\ H^+} \underset{\text{hemi-acetal}}{R.CH(OH)(OR')} \xrightleftharpoons{R'OH,\ H^+} \underset{\text{acetal}}{R.CH(OR')(OR')}$$

Write an equation of this type for the reaction between hexanal, methanol, and a little anhydrous hydrogen chloride.

- -

A24

$CH_3.CH(OH).CN$ $p\text{-}CH_3.C_6H_4.CO.CH(OH).C_6H_4.CH_3\text{-}p$

—CO.CH(OH)— $CH_3.CH_2.CH_2.CH(OH).CN$

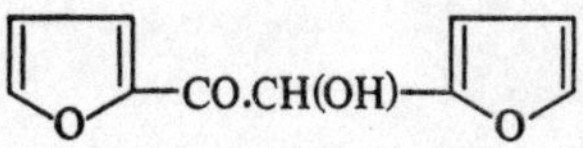

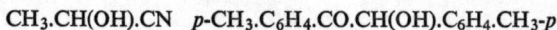

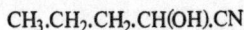

26 Can you formulate a similar reaction between acetone and ethylene glycol ($HOCH_2.CH_2OH$)? Hint: one molecule of the dihydric alcohol is able to replace two molecules of monohydric alcohol.

- -

A25

$CH_3.CH_2.CH_2.CH_2.CH_2.CHO$ will be abbreviated to $C_5H_{11}.CHO$

$$C_5H_{11}.CHO \; \underset{\longleftarrow}{\overset{MeOH,\ H^+}{\longrightarrow}} \; C_5H_{11}.CH \overset{OH}{\underset{OMe}{}} \; \underset{\longleftarrow}{\overset{MeOH,\ H^+}{\longrightarrow}}$$

$$C_5H_{11}.CH \overset{OMe}{\underset{OMe}{}}$$

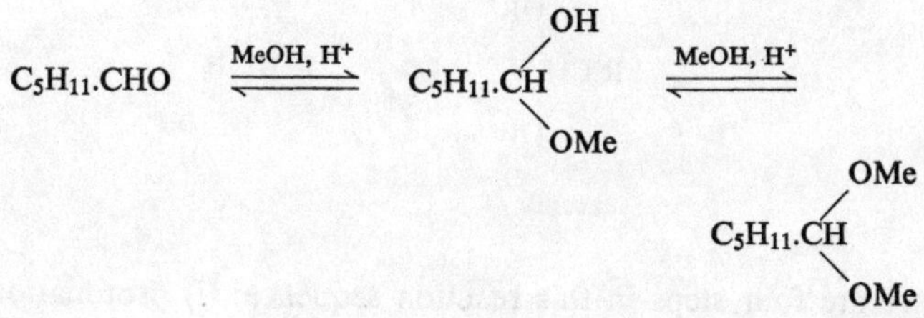

27 Hemiacetal formation is easily understood on the basis of the addition reaction already discussed. This is, however, an acid-catalysed reaction. The aldehyde is protonated giving a more reactive species before it is attacked by an alcohol molecule *behaving as a nucleophilic reagent*. The hemi-acetal results from the protonated species by loss of a proton.

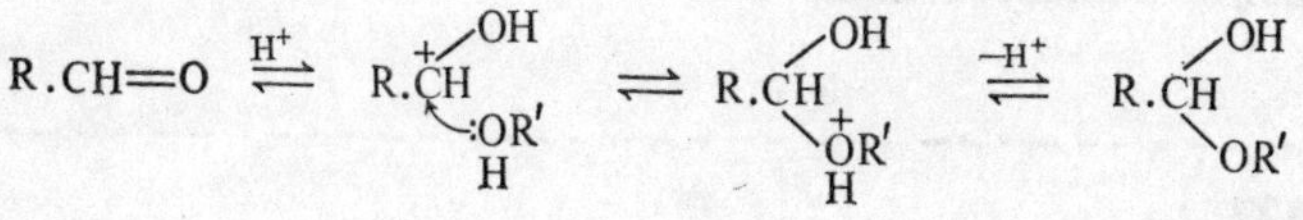

aldehyde   hemi-acetal

Formulate the conversion of hexanal and methanol into the hemi-acetal in this way.

- -

A26

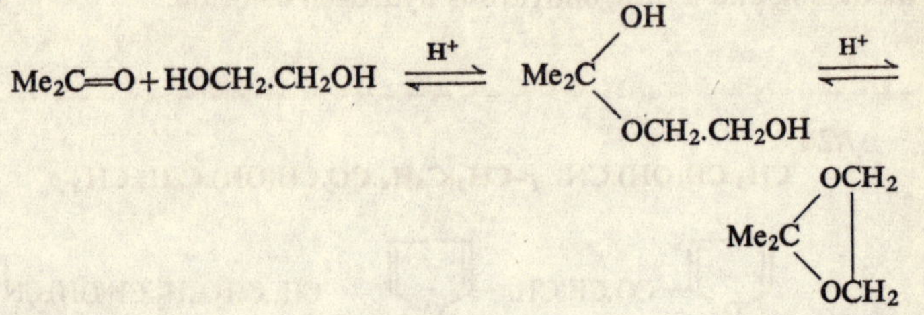

28 The continued reaction of hemi-acetal to acetal is by a sequence of changes which no longer fall into the general pattern discussed so far.

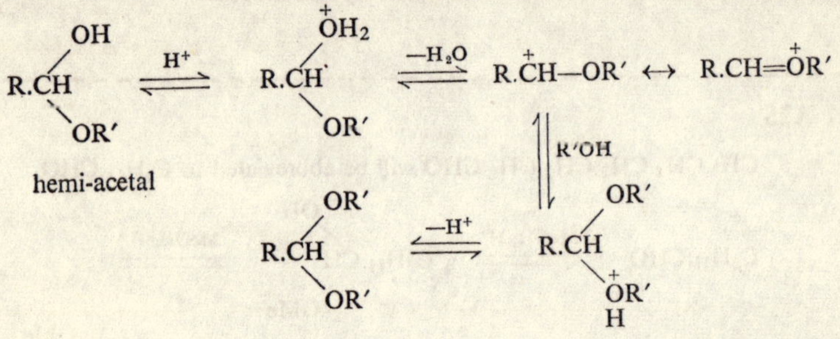

There are four steps in this reaction sequence: (i) protonation of the hemi-acetal on the OH group; (ii) loss of water to give a carbonium ion-oxonium ion resonance species; (iii) attack at the electron-deficient carbon atom in the carbonium ion by a second alcohol molecule behaving as a(n) *electrophilic/nucleophilic* reagent; and (iv) deprotonation to furnish the acetal.

Write the structure of the intermediate carbonium ion-oxonium ion resonance hybrid. Which is the carbonium ion?

Write this sequence for the conversion of hemi-acetal into acetal in the hexanal/methanol reaction.

A27

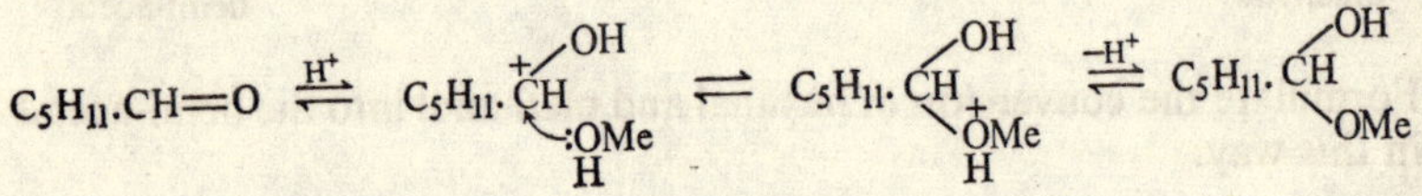

29 Write the mechanism for the complete reaction between acetone and ethylene glycol (for the structure of the products see A26).

A28 nucleophilic $R.\overset{+}{C}H{-}OR' \leftrightarrow R.CH{=}\overset{+}{O}R'$
(carbonium ion) (oxonium ion)

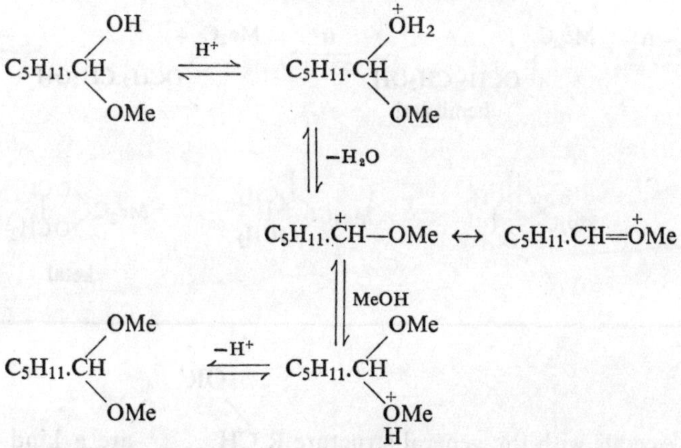

---

30 Notice that in this last example the conversion of hemi-ketal into ketal is an *intramolecular reaction.* Some hydroxy-aldehydes exist mainly as hemi-acetals and this is important in the carbohydrates.

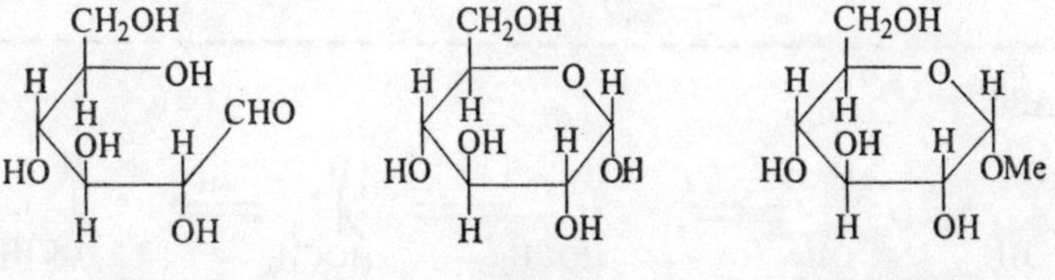

(V) (VI) (VII)

For example, α-D-glucose exists mainly in the cyclic form **(VI)** which is the hemi-acetal of the open-chain hydroxy-aldehyde **(V)**. Interconversion between these occurs very readily: water is a sufficient and excellent catalyst.
Formulate the conversion of **(VI)** into its methylated derivative **(VII)** by reaction with methanolic hydrogen chloride.
Among the structures **(V,)** **(VI)** and **(VII)** there is an aldehyde, hemi-acetal, and acetal. Which is which?

A29

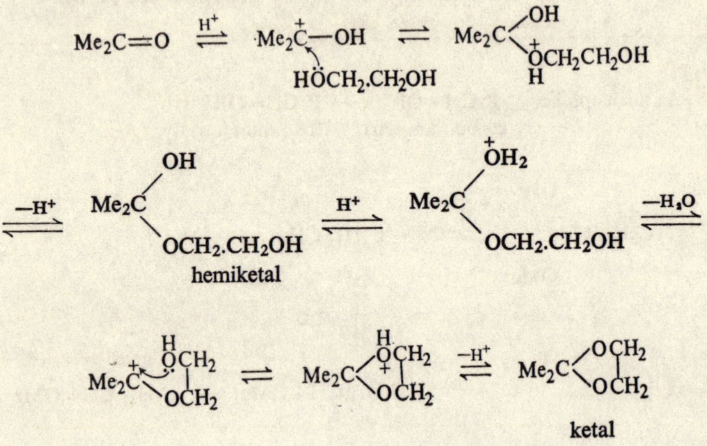

---

31 Acetals with the general structure $R.CH\overset{OR'}{\underset{OR'}{\diagup\diagdown}}$ 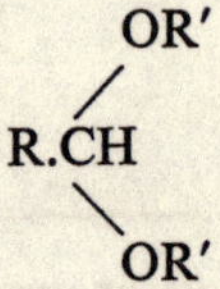 are a kind of *alcohol/ester/aldehyde/ether* with two alkoxy groups (R′O) attached to the same carbon atom.

A30

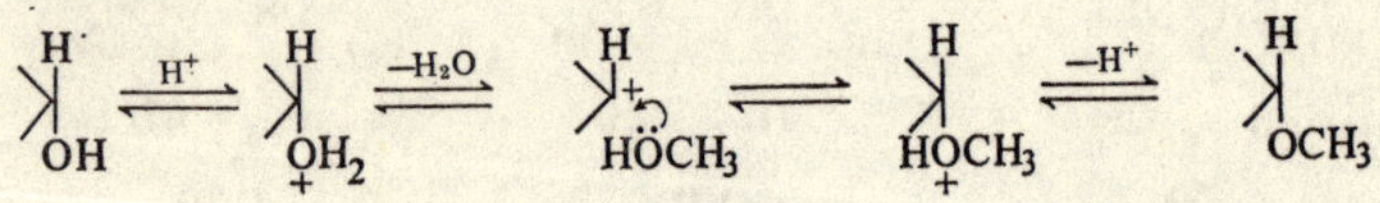

aldehyde (V), hemi-acetal (VI), acetal (VII).

---

32 Like the simpler ethers they are stable to alkali but are cleaved by acids. They differ from the simple ethers in their greater reactivity. The acid cleavage of ethers probably occurs through a carbonium ion thus:

$$ROR' \overset{H^+}{\rightleftharpoons} \underset{H}{R\overset{+}{O}R'} \rightleftharpoons ROH + \overset{+}{R'} \overset{H_2O}{\rightleftharpoons} R'\overset{+}{O}H_2 \rightleftharpoons R'OH + H^+$$

Complete the sequence for an acetal

$$\underset{\substack{\text{acetal}}}{\overset{\displaystyle OR}{\underset{\displaystyle OR}{R'.CH}}} \quad \underset{\xrightarrow{\quad}}{\overset{H^+}{\rightleftharpoons}} \quad ---- \quad \rightleftharpoons \quad ROH+ \quad ---- \quad \leftrightarrow \quad R'.CH{=}\overset{+}{O}R$$

$$\Big\updownarrow H_2O$$
$$\overset{+}{O}H_2$$

$$ROH+ \quad ---- \quad \rightleftharpoons \quad ---- \quad \overset{H^+}{\rightleftharpoons} \quad ---- \quad \overset{-H^+}{\rightleftharpoons} \quad \underset{\substack{\text{hemi-acetal}}}{\overset{\displaystyle R'.CH}{\underset{\displaystyle OR}{}}}$$

$$\Big\updownarrow$$

$$---- \quad \overset{-H^+}{\rightleftharpoons} \quad \underset{\substack{\text{aldehyde}}}{R.CHO}$$

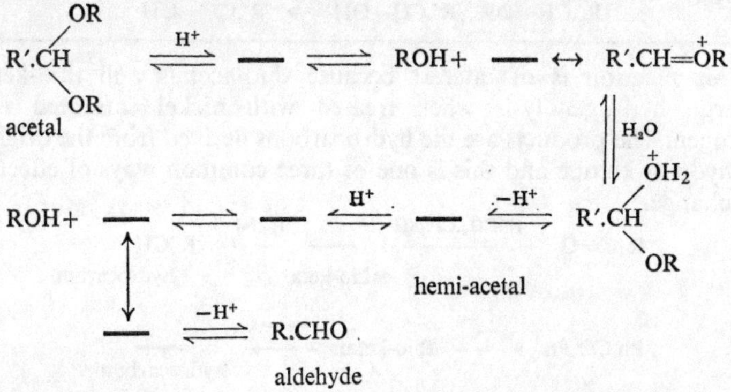

A31 ether.

----

33 The greater reactivity of the acetals can be understood in terms of the stability of the intermediate carbonium ions.

reactions	*intermediate ion*
ether → alcohol	R^+
acetal → hemi-acetal	$R'.\overset{+}{C}H{-}OR \leftrightarrow$ ——
hemi-acetal → aldehyde	—— $\leftrightarrow$ ——

The resonance between carbonium and oxonium ion makes these ions more stable and thus facilitates reactions occurring through their agency.

A32

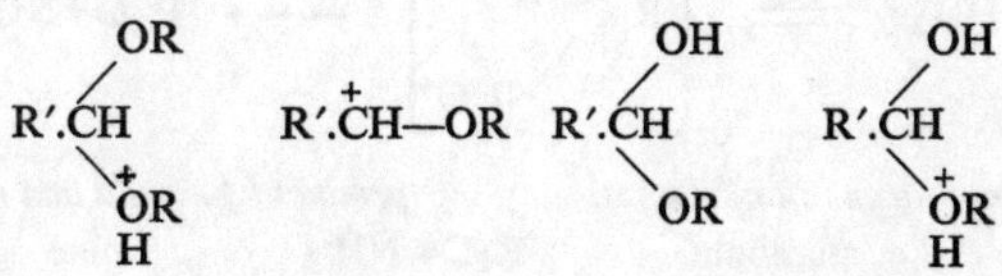

$$\underset{\substack{\overset{+}{O}R \\ H}}{\overset{\displaystyle OR}{R'.CH}} \qquad R'.\overset{+}{C}H{-}OR \qquad \underset{\substack{OR}}{\overset{\displaystyle OH}{R'.CH}} \qquad \underset{\substack{\overset{+}{O}R \\ H}}{\overset{\displaystyle OH}{R'.CH}}$$

$$R\overset{+}{C}H{-}OH \qquad RCH{=}\overset{+}{O}H$$

----

34 Aldehydes and ketones form thio-acetals and thio-ketals by a similar reaction with thiols (RSH),

e.g.
$$\underset{\displaystyle CH_3}{\overset{\displaystyle CH_3}{C}}{=}O \quad + \quad \underset{\displaystyle HSCH_2}{\overset{\displaystyle HSCH_2}{\Big|}} \quad \longrightarrow \quad ---- \quad \text{(thio-ketal)}$$

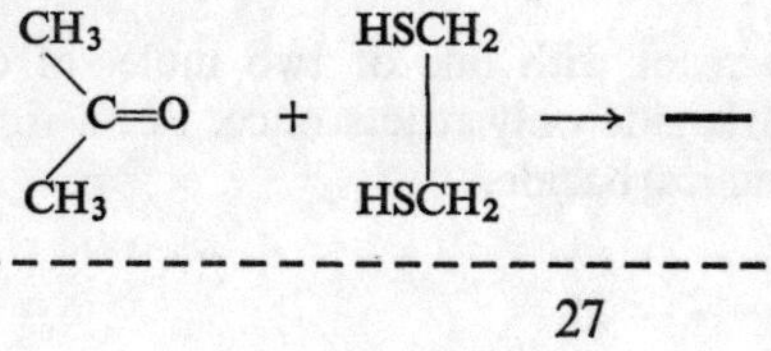

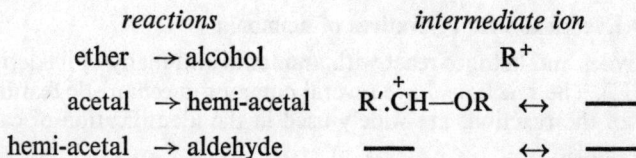

A33 $R'.CH{=}\overset{+}{O}R \quad R'.\overset{+}{C}H{-}OH \leftrightarrow R'.CH{=}\overset{+}{O}H$

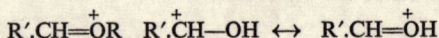

35 This reaction is of interest because thio-acetals and thio-ketals undergo hydrogenolysis when treated with nickel saturated with hydrogen. The products are the hydrocarbons derived from the original aldehyde or ketone and this is one of three common ways of effecting this change.

$$R_2C{=}O \xrightarrow{HSCH_2.CH_2SH} \underset{\text{thio-ketal}}{\text{——}} \xrightarrow{H_2,Ni} \underset{\text{hydrocarbon}}{R_2CH_2}$$

e.g. $$Ph.CO.Ph \longrightarrow \text{thio-ketal} \xrightarrow{\text{——}} \underset{\text{hydrocarbon}}{\text{——}}$$

A34

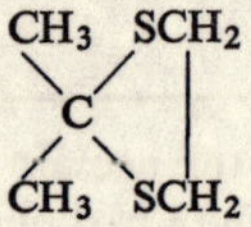

---

36 (v) Reactions with derivatives of ammonia

Aldehydes and ketones react with ammonia and many of its derivatives (GNH_2). The reactions have several common mechanistic features and some of the reactions are widely used in the identification of carbonyl compounds.

Reaction occurs through an intermediate aminohydrin (hydroxy-amine) which is usually dehydrated under the reaction conditions to give the final product.

$$R_2C{=}O + GNH_2 \rightleftharpoons \left[R_2C\begin{matrix} OH \\ NHG \end{matrix} \right] \longrightarrow R_2C{=}NG + H_2O$$

reagent (formula and name)		*product (formula and name)*	
NH_3	ammonia	$R_2C{=}NH$	imine
$R'NH_2$	amine	$R_2C{=}NR'$	imine (Schiff's base)
NH_2NH_2	hydrazine	{ ——	hydrazone
		{ $R_2C{=}NN{=}CR_2$	azine
$PhNHNH_2$	phenylhydrazine	——	phenylhydrazone
NH_2OH	hydroxylamine	——	oxime
$NH_2CONHNH_2$	semicarbazide	$R_2C{=}NNHCONH_2$	semicarbazone

Notice that hydrazine can react with one or two moles of carbonyl compound but that semicarbazide only reacts once. Put a ring round the active NH_2 group in semicarbazide.

$$R_2C \underset{SCH_2}{\overset{SCH_2}{\big<}} \qquad H_2, Ni \quad Ph.CH_2.Ph$$

---

37 The products with ammonia are generally unstable and enter into further complex reactions. Hydrazine is of limited value because it can react in two ways. The others are important.

Complete the following:

$CH_3.CHO + NH_2OH$ (hydroxylamine) → —— (oxime)

$Ph.CHO$ + $PhNH_2$ (——) → —— (imine or —— base)

$Et.CHO$ + —— (——) → $Et.CH:NNHCONH_2$ (——)

—— + $PhNHNH_2$ (——) → $Me_2C=NNHPh$ (——)

A36

$·R_2C=NNH_2$ $R_2C=NNHPh$ $R_2C=NOH$ $NH_2CONH\boxed{NH_2}$

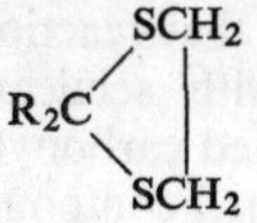

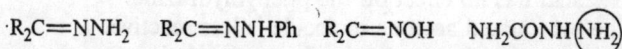

38 The reaction with phenylhydrazine and with semicarbazide is usually catalysed by acid. The protonated aldehyde or ketone is *more/less* reactive. The reaction then becomes:

$$R_2C=O \overset{H^+}{\rightleftharpoons} \begin{array}{c} R_2C=\overset{+}{O}H \\ \updownarrow \\ R_2\overset{+}{C}-OH \\ \underset{NH_2G}{} \end{array} \rightleftharpoons R_2C\underset{\overset{+}{N}H_2G}{\overset{OH}{\big<}} \overset{-H^+}{\rightleftharpoons} R_2C\underset{NHG}{\overset{OH}{\big<}} \overset{-H_2O}{\longrightarrow} R_2C=NG$$

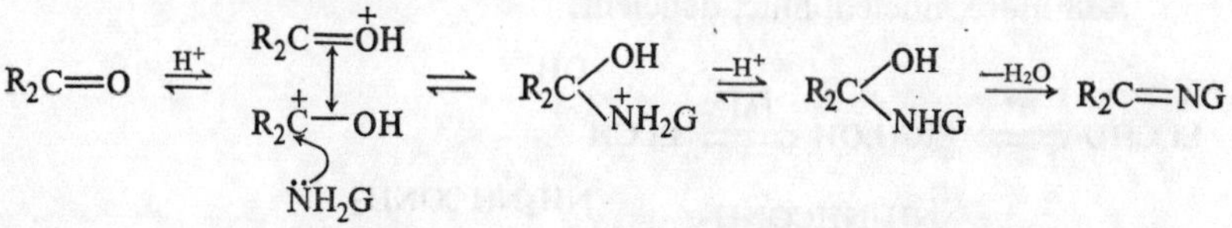

In the step which involves reaction between amine and carbonyl compound, the amine behaves as a(n) *electrophilic/nucleophilic* reagent attacking an electron-*rich/deficient* carbonyl carbon atom. This reaction is an addition process followed by elimination (of water). Formulate the reactions between $Et.CHO$ and $NH_2CONHNH_2$ and between Me_2CO and $PhNHNH_2$ and label clearly the addition and elimination stages in your sequence.

$CH_3.CH=NOH$

aniline, $Ph.CH=NPh$ Schiff's,

$NH_2CONHNH_2$ semicarbazide, semicarbazone,

$Me_2C=O$ phenylhydrazine, phenylhydrazone.

---

39 It was stated in the previous frame that reaction of carbonyl compounds with phenylhydrazine and with semicarbazide is an acid-catalysed process and that the protonated carbonyl compound is more reactive than the non-protonated carbonyl compound. This is true: but the acid can also protonate the basic NH_2 reagents to give compounds which no longer have a lone pair and are, therefore, no longer nucleophilic reagents.

$$R_2C{=}O + H^+ \rightleftharpoons R_2C{=}\overset{+}{O}H \leftrightarrow R_2\overset{+}{C}{-}OH \quad \text{(more reactive)}$$

$$G\ddot{N}H_2 + H^+ \rightleftharpoons G\overset{+}{N}H_3 \quad \text{(unreactive)}$$

There is a delicate balance between these two factors and the amount of acid used to catalyse this reaction must be controlled carefully. Designate each of the following statements as true or false:

(i) Reaction between acetone and phenylhydrazine is usually acid catalysed.
(ii) The acid protonates the carbonyl compound.
(iii) The acid has no effect on the phenylhydrazine.
(iv) The protonated acetone molecule is less reactive.
(v) The protonated phenylhydrazine molecule is less reactive.
(vi) The catalytic effect of acid on this reaction is independent of the amount of acid used.

A38 more, nucleophilic, deficient,

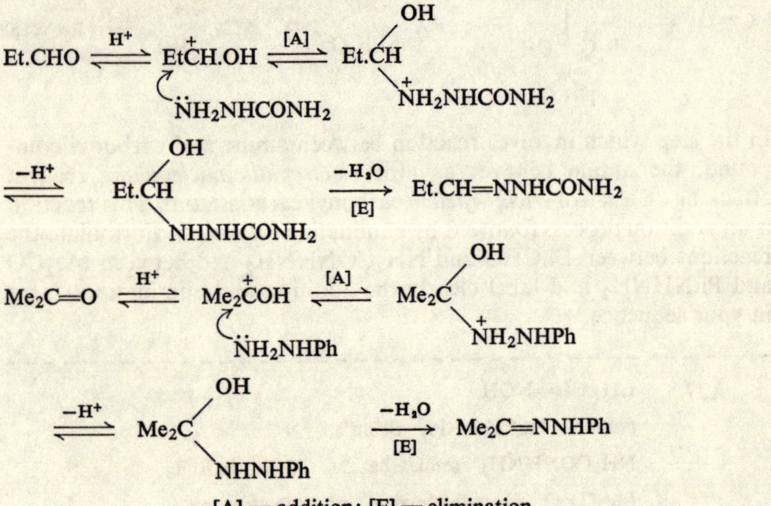

[A] = addition; [E] = elimination

40 Reaction of carbonyl compounds with NH_2OH (name ——)
produces an ——

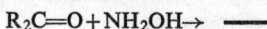

$$R_2C=O + NH_2OH \rightarrow \text{——}$$

This is usually carried out in alkaline solution, *i.e.* it is catalysed by
bases. The basic catalyst converts the NH_2OH into a more reactive
nucleophilic/electrophilic agent by removing a proton. The resulting
anion is tautomeric (not resonance) and it is the $\bar{N}HOH$ species which
attacks the carbonyl compound.

$$NH_2OH + \bar{O}H \rightarrow NH_2\bar{O} \rightleftharpoons \bar{N}HOH$$

$$R_2\overset{\overset{\displaystyle O}{\|}}{C} \curvearrowleft \bar{N}HOH \longrightarrow \text{——} \xrightarrow[]{H^+} \text{——} \xrightarrow[]{-H_2O} \text{——}$$

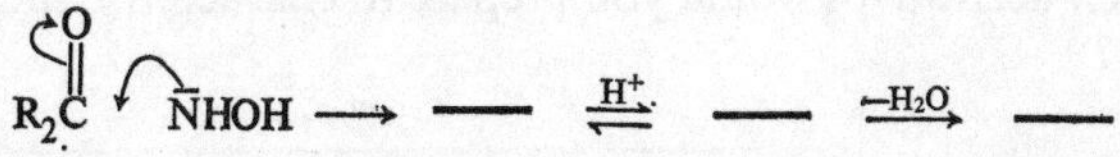

A39 (i) true; (ii) true; (iii) false; (iv) false; (v) true; (vi) false.

---

41 Aldehydes and ketones react with ammonia and its derivatives such
as —— (give names and formulae of at least four). The products contain a new $C=N$ linkage and result from an addition reaction followed
by —— of a water molecule.
Sometimes the reactions are acid-catalysed, as with —— (name two).
The reaction with —— is base-catalysed.
Acid catalysts increase the reactivity of —— by protonation but also
increase/decrease the reactivity of the nucleophilic reagent. Basic
catalysts increase the reactivity of the *carbonyl compound/nucleophilic
reagent*.

- -

A40 hydroxylamine, oxime, $R_2C=NOH$, nucleophilic,

$$R_2C\overset{O^-}{\underset{NHOH}{\diagup}} \qquad R_2C\overset{OH}{\underset{NHOH}{\diagup}} \qquad R_2C=NOH$$

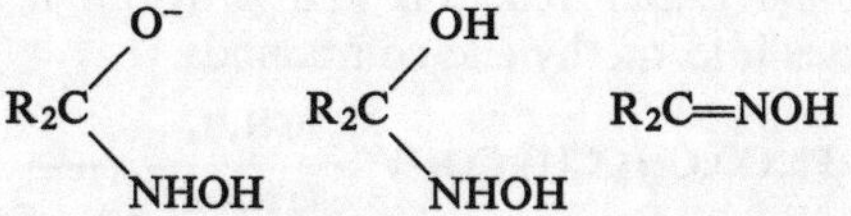

42 These reactions are of great value in the identification of aldehydes
and ketones because in this way liquids are easily converted to crystal-
line derivatives which can be purified and identified by melting point.
Write equations for the conversion of benzaldehyde (Ph.CHO) into its
oxime, phenylhydrazone, and semicarbazone and in each case show the
type of catalyst (H^+ or $\bar{O}H$) you would use (a full mechanistic sequence
is not required).

- -

A41 four from: amines (RNH_2), hydrazine (NH_2NH_2), phenylhydrazine ($PhNHNH_2$), hydroxylamine (NH_2OH), semicarbazide ($NH_2CONHNH_2$), elimination;
phenylhydrazine and semicarbazide (also amines and hydrazines); hydroxlamine;
the carbonyl compound (aldehydes and ketones), decrease, nucleophilic reagent.

---

43 The melting points of the 2,4-dinitrophenylhydrazone, oxime, and semicarbazone of three closely related aldehydes are given in the table below. Which derivatives would you prepare to characterise these three compounds?

	m.p.		
aldehyde	2,4-*dinitrophenylhydrazone*	*oxime*	*semicarbazone*
m-chlorobenzaldehyde	255°	70°	228°
m-bromobenzaldehyde	257	71	205
o-chlorobenzaldehyde	206	75	225

A42

$$Ph.CHO + NH_2OH \xrightarrow{\bar{O}H} Ph.CH{=}NOH$$

$$Ph.CHO + PhNHNH_2 \xrightarrow{H^+} Ph.CH{=}NNHPh$$

$$Ph.CHO + NH_2CONHNH_2 \xrightarrow{H^+} Ph.CH{=}NNHCONH_2$$

---

44 When hydrazones are treated with strong alkali they are reduced to hydrocarbons.

$$\underset{\text{ketone}}{R_2C{=}O} \xrightarrow{N_2H_4} \underset{\text{hydrazone}}{\text{——}} \longrightarrow \underset{\text{hydrocarbon}}{R_2CH_2}$$

This is the Wolff-Kishner reaction and provides a second way of converting ketones into methylene compounds.

$$Ph.CO.CH_2.CH_2.CO_2H \xrightarrow[\text{(ii) NaOH}]{\text{(i) } N_2H_4} \text{——}$$

This change can also be effected by hydrogenolysis of —— using ——. (If you have forgotten, consult frame 35).

A43 The oximes are too close in melting point to be very suitable. The semicarbazone should easily distinguish *m*-bromobenzaldehyde from the other two and the 2,4-dinitrophenylhydrazone should distinguish *o*-chlorobenzaldehyde from the others. The two together would allow each of the three aldehydes to be identified.

45 (vi) Addition of Grignard reagents and other organometallic compounds

Grignard reagents (RMgX) and other organometallic compounds (RNa or RLi) react additively with aldehydes and ketones. This is not surprising for the metal derivatives are nucleophilic reagents.

$$\underset{(a)}{R.C \equiv C} \qquad \underset{(a)}{Na} \qquad \qquad \underset{(b) \ (b)}{R—MgX}$$

Insert ($+$) and ($-$) as required at (a) and δ^+ and δ^- at (b).

A44 $R_2C=NNH_2$ $\qquad\qquad\qquad$ $Ph.CH_2.CH_2.CH_2.CO_2H$
thio-ketals, nickel saturated with hydrogen.

---

46 Thus we have:

$$\underset{\underset{\underset{R'—MgX}{\delta^- \ \delta^+}}{R}}{\overset{\overset{R}{|}}{\underset{|}{\overset{\delta^+ \ \delta^-}{C=O}}}} \longrightarrow \underset{\underset{R' \ MgX}{}}{\overset{R}{\underset{R}{C—O}}} \xrightarrow{\text{dil. acid}} \underset{R'}{\overset{R}{R—C—OH}}$$

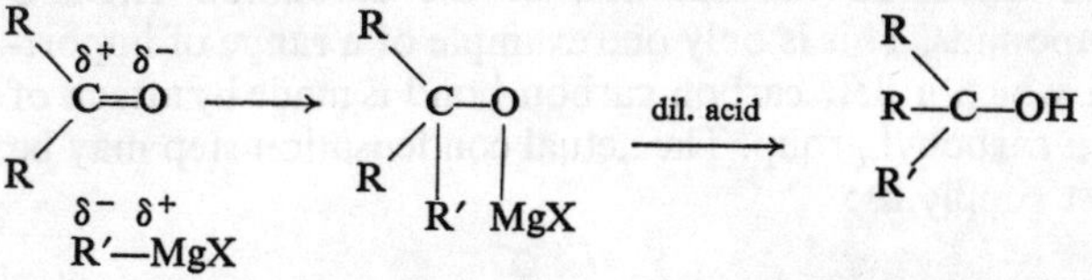

Write a similar sequence for reaction of R.CHO with R'MgX.
We see that ketones give *primary/secondary/tertiary* alcohols and aldehydes give —— alcohols, except formaldehyde which furnishes a —— alcohol.

A45
$$\underset{}{R.C \equiv \overset{-}{C}} \ \overset{+}{Na} \qquad\qquad \underset{}{\overset{\delta^- \ \delta^+}{R—MgX}}$$

---

47 The reaction follows a similar course with the more extensively polarised sodium and lithium compounds:

$$R'.C \equiv \overset{\frown}{\overset{-}{C}} \ \overset{\overset{O}{\|}}{C}R_2 \ \rightleftharpoons \ R'.C \equiv C.\overset{\overset{O^-}{|}}{C}R_2 \xrightarrow{\text{dil. acid}} R'.C \equiv C.C(OH)R_2$$

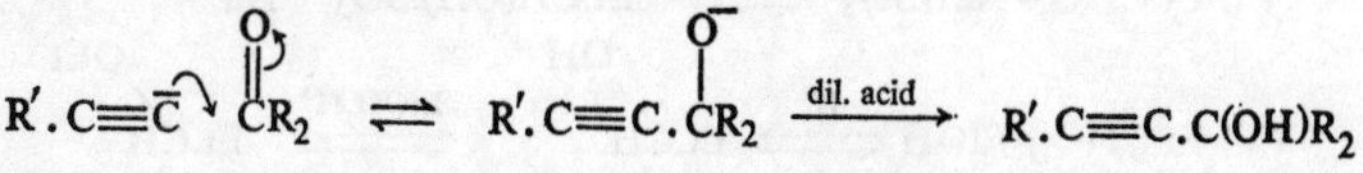

This is a(n) *addition/substitution* reaction. The nucleophilic reagent is a *carbanion/carbonium* ion.

A46 $\underset{\underset{\text{tertiary, secondary, primary.}}{\underset{R'—MgX}{\delta^- \ \delta^+}}}{\overset{\delta^+ \ \ \delta^-}{R.CH=O}} \longrightarrow \underset{R' \ MgX}{R.CH—O} \xrightarrow{\text{dil. acid}} \underset{R'}{\overset{R}{CHOH}}$

C

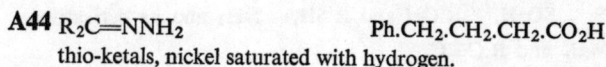

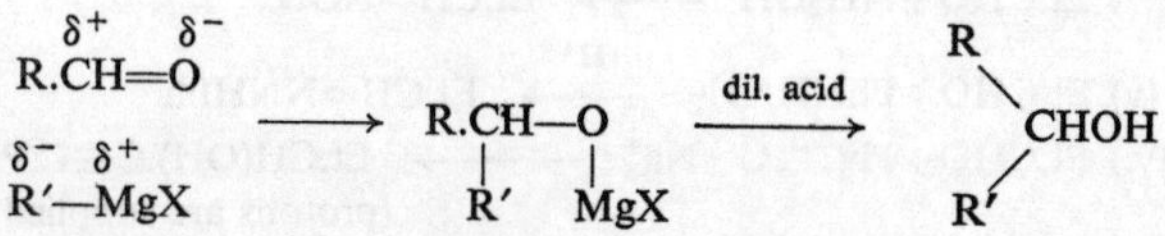

48 We have now discussed several types of nucleophilic reagents which attack aldehydes and ketones. List the ones you can remember.

– –

A47 addition,   carbanion.

49 Formulate the reactions occurring between propionaldehyde and (i) HCN, (ii) NaHSO$_3$, (iii) EtOH and H$^+$, (iv) NH$_2$OH and alkali, (v) PhNHNH$_2$, (vi) Ph.C≡CH and NaNH$_2$. There is no need to write out the full mechanism unless it helps you to formulate the product.

– –

A48 $\overline{C}N$, $\overline{S}O_3H$,   R'OH and R'SH,   NH$_3$ and its derivatives, RMgX and R.C≡$\overline{C}$.

50 (vii) Addition of carbanions

In frame 47 we discussed the addition of the carbanion R.C≡$\overline{C}$ to carbonyl compounds. This is only one example of a range of important reactions in which a new carbon-carbon bond is made by attack of a carbanion on a carbonyl group. The actual condensation step may be represented most simply as:

$$R_3\overline{C} \overset{\curvearrowright}{\underset{}{}} \overset{O}{\underset{}{\overset{\parallel}{C}R_2}} \rightleftharpoons R_3C—\overset{O^-}{\underset{}{C}R_2}$$

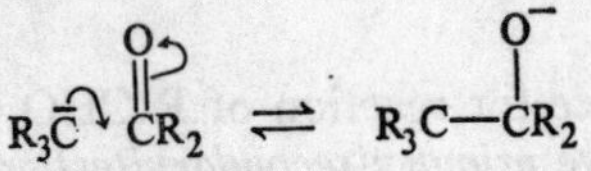

Since the carbanion is often produced from a compound which contains an active methylene group these two reactants can be distinguished as the *methylene component* and the *carbonyl component*.

Before proceeding further with this reaction we must learn something about carbanions.

Continue with frame 51.

– –

A49 (i) Et.CHO + HCN $\rightleftharpoons$ Et.CH(OH).CN

(ii) Et.CHO + NaHSO$_3$ $\rightleftharpoons$ Et.CH(OH).SO$_3^-$ Na$^+$

(iii) Et.CHO + EtOH $\overset{H^+}{\rightleftharpoons}$ Et.$\overset{OH}{\underset{OEt}{CH}}$ $\overset{EtOH.H^+}{\rightleftharpoons}$ Et.$\overset{OEt}{\underset{OEt}{CH}}$

hemi-acetal   acetal

(iv) Et.CHO + NH$_2$OH $\overset{alkali}{\longrightarrow}$ Et.CH=NOH

(v) Et.CHO + PhNHNH$_2$ $\overset{H^+}{\longrightarrow}$ Et.CH=NNHPh

(vi) Et.CHO + PhC≡$\overline{C}$ Na$^+$ $\longrightarrow$ Et.CH(OH).C≡C.Ph
(protons are supplied when working up the product)

51 Carbanions are not reagents which can be stored for use. They must be prepared as required and they are so reactive that they are usually made *in situ*, *i.e.* in the presence of the carbonyl compound with which they are to react.

Carbanions are made from appropriate C—H compounds by proton removal with the help of a base such as $\bar{O}H$ or $\bar{O}Et$,

i.e.

$$R_3CH+\bar{O}H \rightleftharpoons \underset{\text{carbanion}}{\text{——}} +H_2O$$

But not every compound will release a carbon-bound proton in this way. This activity is confined to molecules in which one or more of the R groups contains an electron-withdrawing group *adjacent* to the carbon atom which is to lose its proton. The electron-withdrawing group stabilises the carbanion through delocalisation of the negative charge (resonance).

The important electron-withdrawing groups include NO_2, CN, C=O, CO_2R, and C=C, and it is particularly important for our present consideration that the carbonyl group is in this list. This means that *carbonyl compounds can act both as methylene and carbonyl component* and are therefore capable of self-condensation. Using the list given above put a ring round any CH_3, CH_2, or CH group in the following structures which can form a carbanion and write the formula of the carbanion.

$CH_3.CO.CH_3$ $\quad$ CH_3NO_2 $\quad$ $CH_3.CH_2OH$ $\quad$ $CH_3.CO_2Et$

$CH_3.CH_2NH_2$ $\quad$ $CH_3.CH_2OCH_2.CH_3$ $\quad$ $CH_3.CH_2.CHO$

---

52 Did you get the last one right? It is important to remember that the electron-withdrawing group exerts its special influence only on the *adjacent* CH, CH_2, or CH_3 group.

It was stated in frame 51 that these electron-withdrawing groups stabilise the carbanion (and therefore increase the chance of it being formed) by resonance or delocalisation.

So we can write:

$$CH_3—\bar{C}H—CH=O \leftrightarrow CH_3—CH=CH—\bar{O}$$

Try to write the resonance structures for the three remaining carbanions in the answer to frame 51.

A51 $R_3\bar{C}$

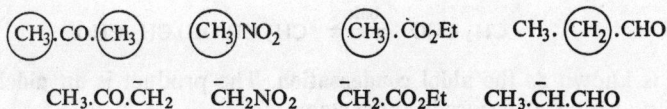

(CH_3).CO.(CH_3) $\quad$ (CH_3)NO_2 $\quad$ (CH_3).CO_2Et $\quad$ CH_3.(CH_2).CHO

$CH_3.CO.\bar{C}H_2$ $\quad$ $\bar{C}H_2NO_2$ $\quad$ $\bar{C}H_2.CO_2Et$ $\quad$ $CH_3.\bar{C}H.CHO$

53 Proton-removal, leading to a carbanion, is possible with one activating group but occurs more easily with two or three activating groups. Divide the following compounds into those with no, one, two, and three activating groups and write all the resonance possibilities for the carbanions derived from those marked with an asterisk (*).

(i) $CH_3.CO.CH_3$ (ii) $CH_3.CO_2Et$ (iii) $CH_3.CH_2.CH_2OH$ (iv) $Me_3C.CHO$

(v) $CH_3.CO.CH_2.CO_2Et*$ (vi) $EtO_2C.CH_2.CN$ (vii) CH_3NO_2

(viii) $(CH_3.CO)_3CH$ (ix) $EtO_2C.CH_2.CO_2Et$ (x) $CH_3.CN*$ (xi) CH_3NH_2

A52

$$\overset{-}{C}H_2-C=O \leftrightarrow CH_2=C-\overset{-}{O}$$
$$\quad\quad\quad | \quad\quad\quad\quad\quad | $$
$$\quad\quad CH_3 \quad\quad\quad\quad CH_3$$

$$\overset{-}{C}H_2-\overset{+}{N}=O \leftrightarrow CH_2=\overset{+}{N}-\overset{-}{O}$$
$$\quad\quad\quad | \quad\quad\quad\quad\quad | $$
$$\quad\quad O^- \quad\quad\quad\quad O^-$$

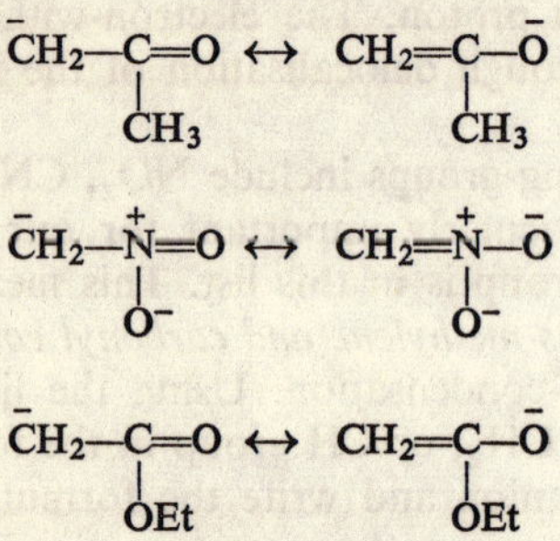

(examine this carefully if your answer is wrong)

$$\overset{-}{C}H_2-C=O \leftrightarrow CH_2=C-\overset{-}{O}$$
$$\quad\quad\quad | \quad\quad\quad\quad\quad | $$
$$\quad\quad OEt \quad\quad\quad\quad OEt$$

---

54 Since a carbanion can be produced from so many different types of compounds and the carbonyl component can be aldehyde, ketone, or several of the carboxylic acid derivatives, the condensation reactions between carbanions and carbonyl compounds cover a wide area of chemistry.

The reaction can be considered to occur in the following stages.

(i) Production of carbanion by interaction of basic catalyst with the activated methylene component.

(ii) Addition reaction between carbanion and carbonyl compound involving nucleophilic attack and (usually) subsequent protonation.

(iii) Further reaction of the addition product.

The base-catalysed self-condensation of aldehydes and ketones provides a simple example,

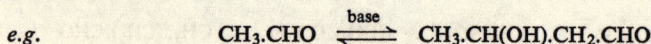

e.g. $CH_3.CHO \overset{\text{base}}{\rightleftharpoons} CH_3.CH(OH).CH_2.CHO$

This is known as the **aldol condensation**. The product is an **ald**ehyde and an **alco**hol. It occurs in three stages:

formation of carbanion $CH_3.CHO + \bar{O}H \rightleftharpoons$ —— $+ H_2O$

reaction between carbanion and carbonyl compound

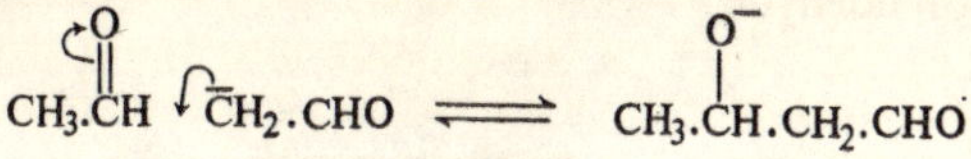

protonation $CH_3.\overset{O^-}{\overset{|}{C}}H.CH_2.CHO + H_2O \rightleftharpoons$ —— $+ \bar{O}H$

A53 no activating group (iii) (xi),
one activating group (i) (ii) (vii) (x),
two activating groups (v) (vi) (ix),
three activating groups (viii),
(iv) contains one activating group but has no hydrogen atom on the adjacent carbon atom and therefore cannot form a carbanion.

$$O{=}\overset{CH_3}{\overset{|}{C}}{-}\bar{C}H{-}\overset{OEt}{\overset{|}{C}}{=}O \leftrightarrow \bar{O}{-}\overset{CH_3}{\overset{|}{C}}{=}CH{-}\overset{OEt}{\overset{|}{C}}{=}O \leftrightarrow O{=}\overset{CH_3}{\overset{|}{C}}{-}CH{=}\overset{OEt}{\overset{|}{C}}{-}\bar{O}$$

$$\bar{C}H_2{-}C{\equiv}N \leftrightarrow CH_2{=}C{=}\bar{N}$$

---

55 Write three equations like those in frame 54 for the self-condensation of propionaldehyde in the presence of alkali.

A54 $\bar{C}H_2.CHO$ $CH_3.CH(OH).CH_2.CHO$

---

56 If you got that right move on to frame 59.
Did you finish up with $CH_3.CH_2.CH(OH).CH_2.CH_2.CHO$? You forgot that the C=O group activates only the *adjacent* CH_2 group. What is the correct structure for the carbanion from propionaldehyde?

A55 $CH_3.CH_2.CHO + \bar{O}H \rightleftharpoons CH_3.\bar{C}H.CHO + H_2O$

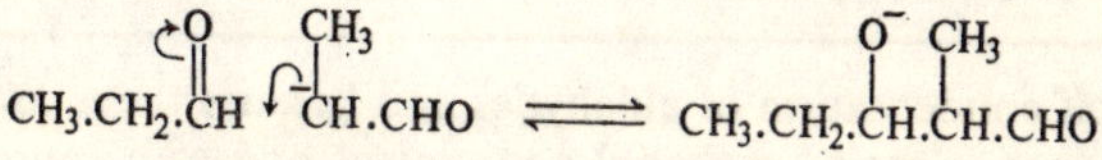

$$CH_3.CH_2.\overset{\bar{O}}{\overset{|}{C}}H.\overset{CH_3}{\overset{|}{C}}H.CHO + H_2O$$

$$\rightleftharpoons CH_3.CH_2.CH(OH).CH(CH_3).CHO + \bar{O}H$$

---

57 Now write the complete reaction sequence for the self-condensation of propionaldehyde.

A56

$$CH_3.\bar{C}H.CHO$$

---

58 Try again with acetone.

A57 See answer to frame 55.

---

59 Under all except the mildest conditions, β-hydroxy-carbonyl compounds are readily dehydrated to αβ-unsaturated carbonyl compounds.

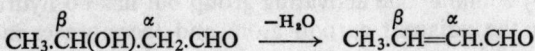

$$CH_3.\overset{\beta}{C}H(OH).\overset{\alpha}{C}H_2.CHO \xrightarrow{-H_2O} CH_3.\overset{\beta}{C}H{=}\overset{\alpha}{C}H.CHO$$

With a basic catalyst this involves the following (showing only the essential atoms):

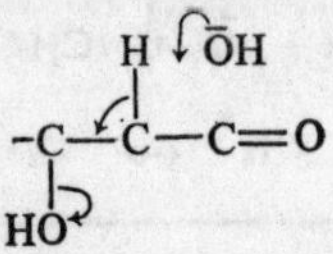

Note that a second proton is now removed from the activated position. It is this activating influence which causes the easy dehydration. Notice too that this dehydration is irreversible.

$$2CH_3.CH_2.CHO \rightleftharpoons CH_3.CH_2.CH(OH).\underset{}{C}H(CH_3).CHO \rightarrow \text{——} + H_2O$$

$$2CH_3.CO.CH_3 \rightleftharpoons \text{——} \rightarrow \text{——} + H_2O$$

A58

$$CH_3.CO.CH_3 + \bar{O}H \rightleftharpoons CH_3.CO.\bar{C}H_2 + H_2O$$

$$(CH_3)_2C{=}O \;\; \bar{C}H_2.CO.CH_3 \rightleftharpoons (CH_3)_2C(O^-).CH_2.CO.CH_3$$

$$(CH_3)_2C(O^-).CH_2.CO.CH_3 + H_2O \rightleftharpoons (CH_3)_2C(OH).CH_2.CO.CH_3 + \bar{O}H$$

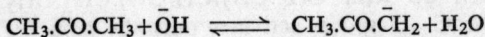

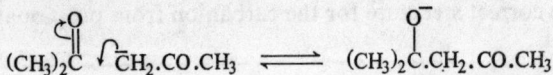

60 In the self condensation of aldehydes and ketones these compounds act in two capacities; as carbonyl component and as a source of the ——. Can you think of any aldehydes or ketones which would not react in this way? (*Clue:* aldehydes and ketones always contain the carbonyl function, could there be conditions when a carbanion could not be formed?)

A59 $CH_3.CH_2.CH{=}C(CH_3).CHO$

$(CH_3)_2\overset{OH}{\overset{|}{C}}.CH_2.CO.CH_3$ $\qquad (CH_3)_2C{=}CH.CO.CH_3$

---

61 This holds for the following. Put a ring round each carbon atom adjacent to a carbonyl group and notice that it has no attached hydrogen atoms.

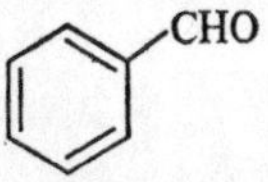

$CH_3{-}\underset{CH_3}{\overset{CH_3}{C}}{-}CHO$ $\qquad HCHO$

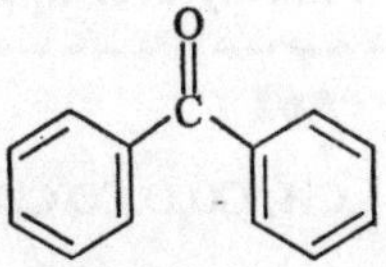

A60 carbanion,
aldehydes and ketones with no H atom on the α-carbon atom cannot form carbanions.

---

62 Though carbonyl compounds of this type cannot undergo self-condensation they can react with carbanions produced from other molecules. Two important applications of this are the **Perkin** reaction and the **Knoevenagel-Doebner** reaction. The former involves the condensation of an aromatic aldehyde with an acid anhydride in the presence of a basic catalyst. This is usually the salt of the carboxylic acid but can be an amine,

e.g. $Ph.CHO + CH_3.CO.O.CO.CH_3 \xrightarrow[\text{acetate}]{\text{sodium}} Ph.CH{=}CH.CO_2H + CH_3.CO_2H$

In this reaction the basic catalyst is $CH_3.COO^-$, the methylene component is acetic anhydride, and the carbonyl component is the aromatic aldehyde. We can now fit this information into the general mechanistic pathway discussed in frame 53.

$CH_3.CO.O.CO.CH_3 + CH_3.COO^- \rightleftharpoons$ ——— + ———
(monovalent carbanion)

$Ph.\overset{O}{\overset{\|}{C}}H$ + ——— $\rightleftharpoons$ ———

——— + $CH_3.COOH \rightleftharpoons CH_3.COO^- + Ph.\overset{OH}{\overset{|}{C}}H.CH_2.CO.O.CO.CH_3$

$Ph.\overset{OH}{\overset{|}{C}}H.CH_2.CO.O.CO.CH_3 \longrightarrow Ph.CH{=}CH.CO_2H + CH_3.CO_2H$

A61

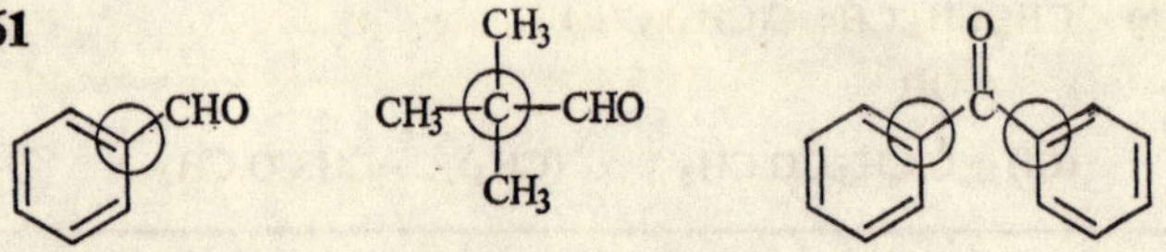

Formaldehyde has no α-carbon atom.

63 Why is this reaction confined almost entirely to *aromatic* aldehydes and hardly ever applied to *aliphatic* aldehydes?

A62

$\bar{C}H_2.CO.O.CO.CH_3 \; + \; CH_3.CO_2H$

$\bar{C}H_2.CO.O.CO.CH_3 \quad Ph.CH(O^-).CH_2.CO.O.CO.CH_3$

$Ph.CH(O^-).CH_2.CO.O.CO.CH_3$

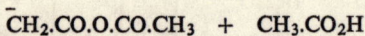

64 In the Knoevenagel-Doebner reaction aromatic aldehydes (carbonyl component) condense with malonic ester or acid (methylene component) in the presence of ammonia or an amine (basic catalyst).

$$Ph.CH{=}C(CO_2Et)_2 \xleftarrow{CH_2(CO_2Et)_2} Ph.CHO \xrightarrow[-CO_2]{CH_2(CO_2H)_2} Ph.CH{=}CH.CO_2H$$

With this information indicate the steps for the conversion of Ph.CHO into $Ph.CH{=}C(CO_2Et)_2$ with RNH_2 as catalyst.

A63 Most aliphatic aldehydes can form a carbanion and so self-condensation would occur as well as the desired Perkin reaction. Aromatic aldehydes cannot form carbanions.

65 (viii) The Cannizzaro reaction

Some aldehydes do not undergo self-condensation with alkali because they cannot form a carbanion. Give two examples.

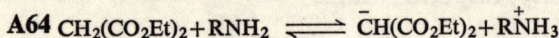

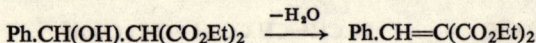

A64 $CH_2(CO_2Et)_2 + RNH_2 \rightleftharpoons \bar{C}H(CO_2Et)_2 + R\overset{+}{N}H_3$

$$Ph.CH{=}O \;\; \bar{C}H(CO_2Et)_2 \rightleftharpoons Ph.CH(O^-).CH(CO_2Et)_2$$

$$Ph.CH(O^-).CH(CO_2Et)_2 + R\overset{+}{N}H_3 \rightleftharpoons Ph.CH(OH).CH(CO_2Et)_2 + RNH_2$$

$$Ph.CH(OH).CH(CO_2Et)_2 \xrightarrow{-H_2O} Ph.CH{=}C(CO_2Et)_2$$

66 With strong alkali these aldehydes undergo a different kind of reaction known as the Cannizzaro reaction. Here is a simple example:

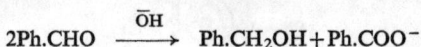

$$2Ph.CHO \xrightarrow{\bar{O}H} Ph.CH_2OH + Ph.COO^-$$

Two molecules of aldehyde are involved in the reaction: one is oxidised to the anion of a carboxylic acid and the other is reduced to a primary alcohol.

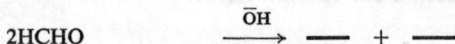

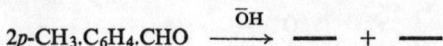

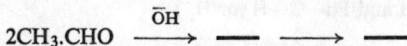

$$2HCHO \xrightarrow{\bar{O}H} \text{——} + \text{——}$$

$$2p\text{-}CH_3.C_6H_4.CHO \xrightarrow{\bar{O}H} \text{——} + \text{——}$$

$$2CH_3.CHO \xrightarrow{\bar{O}H} \text{——} \longrightarrow \text{——}$$

A65 Ph.CHO, $Me_3C.CHO$, any aldehyde which does not have any hydrogen on the adjacent carbon atom, and formaldehyde.

---

67 During reaction there is *transfer of hydride ion* (H^-) from one aldehyde molecule to the other:

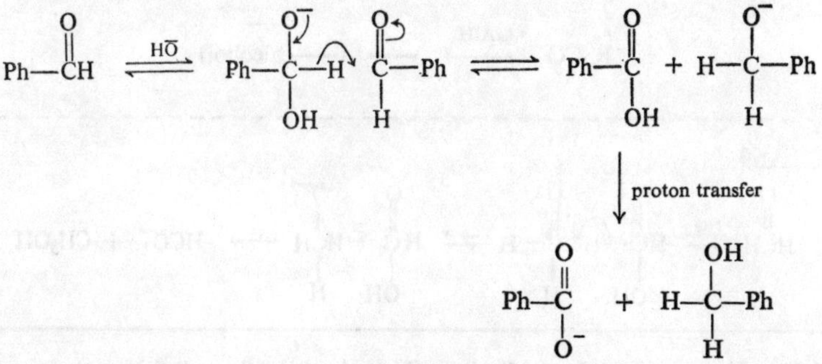

Put a ring round that part of the above formulation which shows the hydride transfer.

A66 $CH_3OH + H.COO^-$ (or H.COOH)

$p\text{-}CH_3.C_6H_4.CH_2OH + p\text{-}CH_3.C_6H_4.COO^-$ (or $p\text{-}CH_3.C_6H_4.COOH$)

$CH_3.CH(OH).CH_2.CHO$ $CH_3.CH{=}CH.CHO$ (hope you missed that trap!)

---

68 Both aldehyde molecules are attacked by nucleophilic reagents at their electron-deficient carbonyl carbon atoms. What are these two nucleophilic reagents?

A67

$$Ph-\underset{\underset{OH}{|}}{C}(H^{\curvearrowright})\underset{\underset{H}{|}}{\overset{O}{C}}-Ph$$

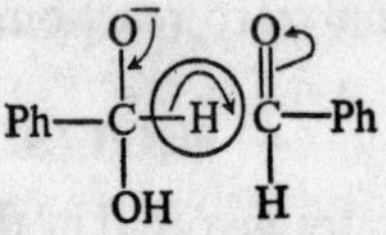

69 Formulate the reaction for formaldehyde.

A68

$$\bar{O}H \text{ and } Ph-\underset{\underset{OH}{|}}{\overset{O^-}{C}}-H \text{ (or } H^-)$$

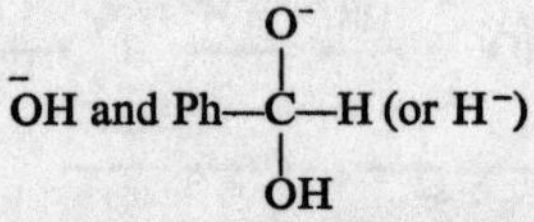

70 (ix) Reduction by metal hydrides

Aldehydes and ketones are reduced by lithium aluminium hydride $(LiAlH_4)$ and sodium borohydride $(NaBH_4)$ to primary and secondary alcohols.

$$RCHO \xrightarrow{LiAlH_4} \quad —— \quad (—— \text{ alcohol})$$

$$R_2CO \xrightarrow{LiAlH_4} \quad —— \quad (—— \text{ alcohol})$$

A69

$$\underset{HCH}{\overset{O}{\parallel}} \rightleftharpoons \overset{HO^-}{\underset{OH}{\overset{O}{HC}}}-H \overset{O}{C}-H \rightleftharpoons \underset{OH}{\overset{O}{HC}} + \underset{H}{\overset{O^-}{HCH}} \longrightarrow HCO_2^- + CH_3OH$$

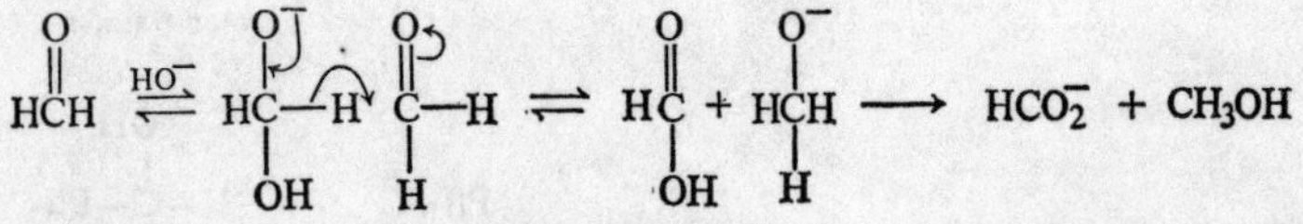

71 In this reaction hydride ion (H^-) is transferred from the metal hydride to the electron-deficient carbon atom and the oxygen reacts with the aluminium or boron compound. The resulting complex is later decomposed by acid.

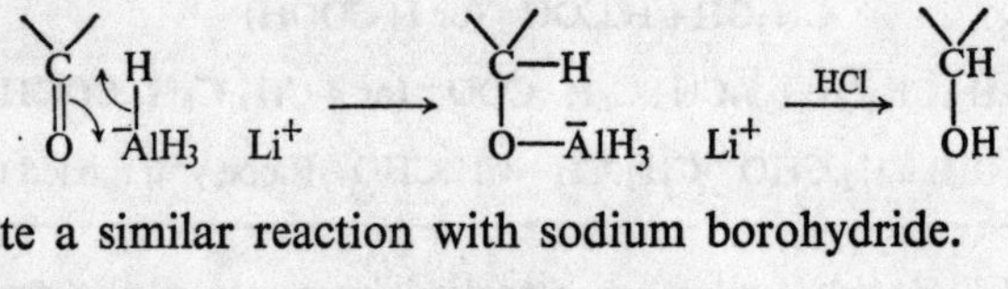

Formulate a similar reaction with sodium borohydride.

A70 RCH_2OH (primary), R_2CHOH (secondary).

72 Reduction occurs in two stages, (i) addition of —— ion from the metal hydride, (ii) addition of *hydride/proton* from the acid. The metal hydride supplies *one/two/three/four* atom(s) of hydrogen per mole of carbonyl compound reduced and therefore one mole of metal hydride ($LiAlH_4$ or $NaBH_4$) should be able to reduce *four/three/two/one* mole(s) of carbonyl compound.

A71

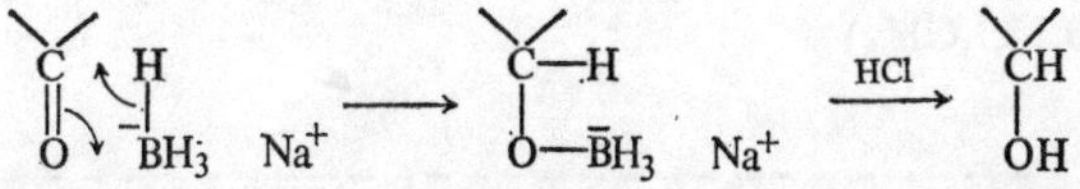

---

73 The reduction of ketones ($R_2C{=}O$) to hydrocarbons (R_2CH_2) can be achieved by the following reactions (if you have difficulties, refer back to frames 35 and 44).

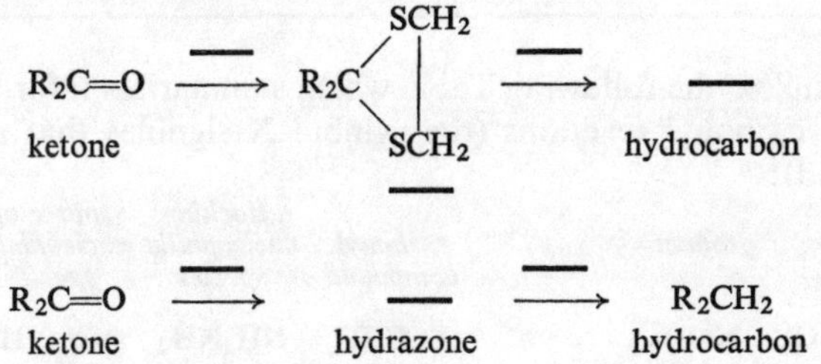

The latter is the W—— K—— reaction.

A72 hydride (H^-), proton (H^+), one, four.

---

74 Test frames

Complete the following sequence which shows the two major pathways of carbonyl reactions:

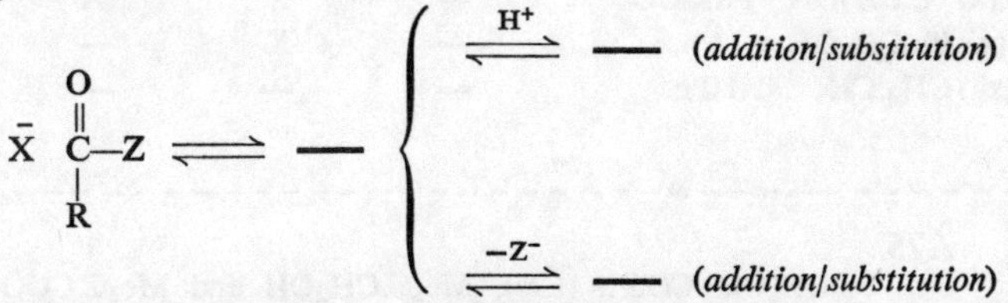

Which is the more common with aldehydes and ketones?

A73

$HSCH_2.CH_2SH$, H^+	thio-ketal	Ni, H_2	R_2CH_2
NH_2NH_2	$R_2C{=}NNH_2$	NaOH	Wolff-Kishner

75 Give the structural formulae of the products of the following reactions. What additional reagents (if any) are required?
benzoin condensation (Ph.CHO), Cannizzaro reaction ($Me_3C.CHO$) aldol condensation (Et.CHO), Perkin reaction (p-$CH_3.C_6H_4.CHO$ and $CH_3.CO.O.CO.CH_3$)

A74 $X{-}\underset{R}{\overset{O^-}{C}}{-}Z$ $X{-}\underset{R}{\overset{OH}{C}}{-}Z$ (addition), $X{-}\underset{R}{\overset{O}{\overset{\|}{C}}}$ (substitution), addition.

76 Complete the following Table which summarises information about several carbonyl reactions (the symbol X signifies that no answer is required):

product	*carbonyl compound*	*attacking nucleophilic species*	*source of nucleophilic species*	*catalyst (if any)*
e.g. $Et.CH{=}NNH_2$	Et.CHO	NH_2NH_2	NH_2NH_2	none
(i) $Ph.CH{=}NPh$	—	—	—	—
(ii) $CH_3.CH(OH).CN$	—	—	—	X
(iii) $Ph.CH(OMe)_2$	—	—	—	—
(iv) Ph.CO.CH(OH).Ph	—	X	X	—
(v) $Et.CH(OH)S\bar{O}_3Na^+$	—	—	—	—
(vi) $Bu^n.C{\equiv}C.CH(OH).Et$	—	—	—	X
(vii) $Me_2C{=}NOH$	—	—	—	—
(viii) $Ph.CH{=}NNHCONH_2$	—	—	—	—
(ix) $CH_3.CH(OH).CH_2.CHO$	—	—	—	—
(x) $Ph.CH_2OH + Ph.COO^-$	—	X	X	—
(xi) $Ph.CH_2OH$	—	X	—	X
(xii) $CH_3.CH{=}NNHPh$	—	—	—	—

A75

Ph.CH(OH).CO.Ph ($\bar{C}N$), $Me_3C.CH_2OH$ and $Me_3C.COO^-$ (NaOH),
$CH_3.CH_2.CH(OH).CH(CH_3).CHO$ (NaOH),
p-$CH_3.C_6H_4.CH{=}CH.CO_2H$ ($CH_3.COONa$).

A76 (i) Ph.CHO, $PhNH_2$, $PhNH_2$, none,
(ii) CH_3.CHO, $\bar{C}N$, NaCN (or KCN) or HCN,
(iii) Ph.CHO, MeOH, MeOH, H^+,
(iv) Ph.CHO(2 mols.), $\bar{C}N$,
(v) Et.CHO, $\bar{S}O_3H$, $NaHSO_3$, none,
(vi) Et.CHO, $Bu^n.C{\equiv}\bar{C}$, $Bu^n.C{\equiv}CNa$ or $Bu^n.C{\equiv}CH + NaNH_2$ or lithium compounds,
(vii) Me_2CO, $\bar{N}HOH$, NH_2OH, NaOH,
(viii) Ph.CHO, $NH_2CONHNH_2$, $NH_2CONHNH_2$, H^+,
(ix) CH_3.CHO, $\bar{C}H_2$.CHO, CH_3.CHO, NaOH,
(x) Ph.CHO(2 mols.), NaOH,
(xi) Ph.CHO, $LiAlH_4$ or $NaBH_4$,
(xii) CH_3.CHO, $PhNHNH_2$, $PhNHNH_2$, H^+,

There are 40 answers in this last test frame. Record the number of correct answers you got.

PART 3. CARBOXYLIC ACIDS AND THEIR DERIVATIVES

77 Nucleophilic reagents attack aldehydes and ketones at the electron-deficient carbonyl carbon atom and this is usually followed by protonation of the anionic intermediate. The overall result is a(n) —— reaction. Sometimes this initial reaction is followed by other processes such as an irreversible dehydration.
Carboxylic acids and their derivatives also react with nucleophilic reagents, but this time the primary attack on electron-deficient carbon is followed by an elimination. Overall, the reaction is —— at a(n) *saturated/unsaturated* carbon atom.

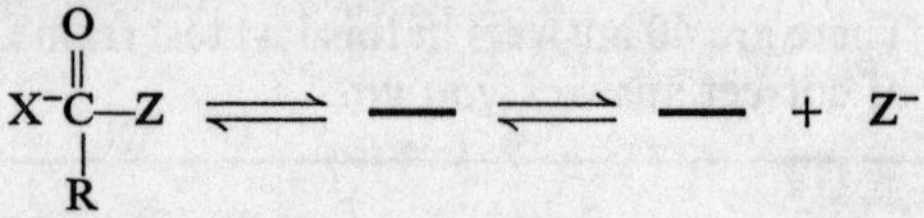

78 In this section we shall be concerned mainly, but not entirely, with the carboxylic acids and four derived series of compounds. Can you formulate and name the following derivatives of acetic acid?

general name	*structure*	*specific name*
carboxylic acid	$CH_3.COOH$	acetic acid
acyl chloride	—	—
anhydride	—	—
(ethyl) ester	—	—
amide	—	—

A77 addition, substitution, unsaturated,

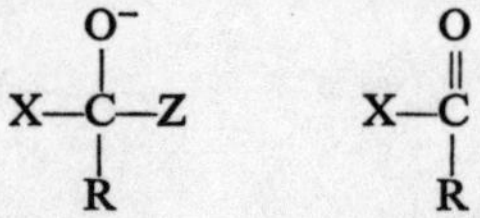

79 All these compounds contain the $CH_3.\overset{O}{\overset{\|}{C}}$ group. This is called an *acetyl* group. The more general structure $R.\overset{O}{\overset{\|}{C}}$ is an acyl group. Associate the following names with the appropriate structure:

phenyl	alkyl	CH_3-	$R-$
propionyl	methyl	C_6H_5-	$C_6H_5.CO-$
benzoyl	acyl	$CH_3.CO-$	$Et.CO-$
acetyl		$R.CO-$	

A78 CH$_3$.COCl (acetyl chloride),   CH$_3$.COO.CO.CH$_3$ (acetic anhydride),
CH$_3$.COOEt (ethyl acetate),   CH$_3$.CONH$_2$ (acetamide).

---

80 If we use the general structure R.COZ to represent the various types of acyl derivatives then we have:

derivative	Z	Z⁻
R.CO$_2$H	OH	⁻OH
R.COCl	—	—
R.COO.COR	—	—
R.COOR′	—	—
R.CONH$_2$	—	—
R.CONHR′	—	—

A79 CH$_3$− (methyl), C$_6$H$_5$− (phenyl), CH$_3$.CO− (acetyl), R− (alkyl), C$_6$H$_5$.CO− (benzoyl), Et.CO− (propionyl), R.CO− (acyl).

---

81 In the reactions we are to consider the group Z will usually be eliminated as Z⁻ (see list in frame 80) and the ease of reaction will depend partly on the stability of Z⁻. The more stable it is the more likely it is to be eliminated.

(most stable) $\underset{\text{group 7}}{\underline{Cl}}$ > $\underset{\text{group 6}}{\underline{\bar{O}CO.R > \bar{O}R > \bar{O}H}}$ > $\underset{\text{group 5}}{\underline{\bar{N}HR > \bar{N}H_2}}$ (least stable)

(groups 5, 6, and 7 refer to the position of N, O, and Cl in the Periodic Table). Some possible reaction intermediates are given below. What two products are possible and which is more likely to be formed?

e.g. HO—$\overset{O^-}{\underset{R}{C}}$—Cl ⟶ HO—$\overset{O}{\overset{\|}{\underset{R}{C}}}$ + Cl⁻ is more likely than H$\bar{O}$ + $\overset{O}{\overset{\|}{\underset{R}{C}}}$—Cl

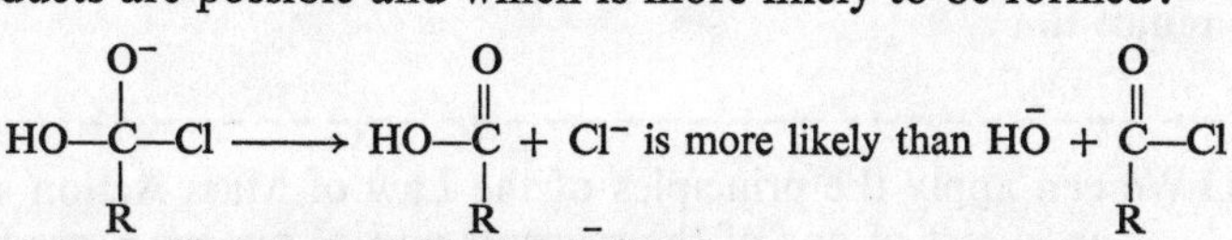

because Cl⁻ is more stable than $\bar{O}$H

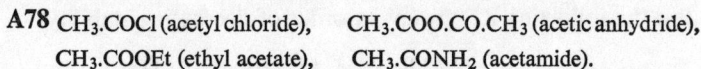

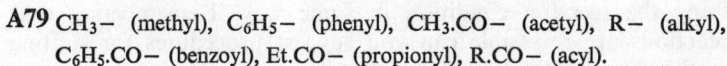

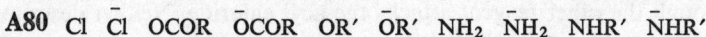

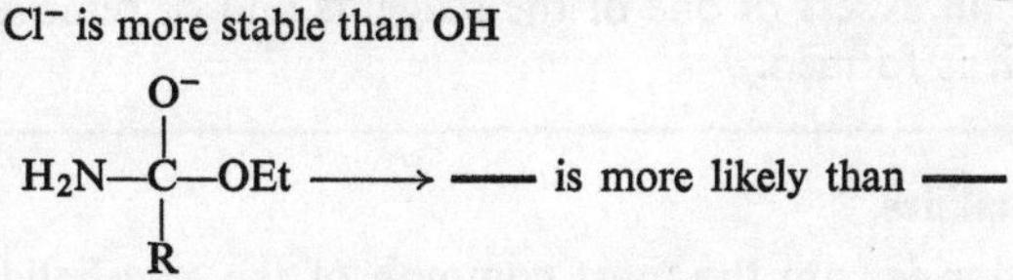

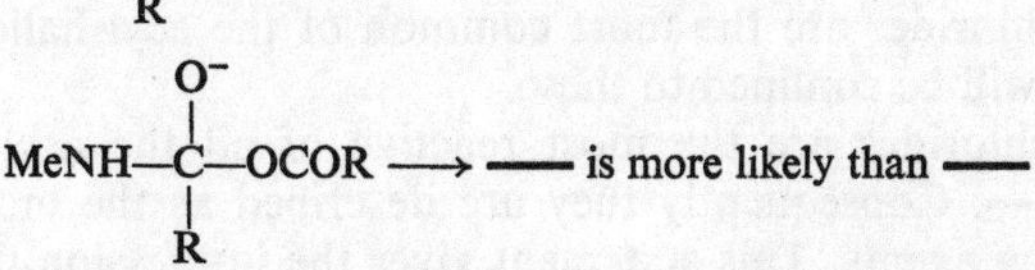

H$_2$N—$\overset{O^-}{\underset{R}{C}}$—OEt ⟶ —— is more likely than ——

MeNH—$\overset{O^-}{\underset{R}{C}}$—OCOR ⟶ —— is more likely than ——

A80 Cl $\bar{C}$l OCOR $\bar{O}$COR OR′ $\bar{O}$R′ NH$_2$ $\bar{N}$H$_2$ NHR′ $\bar{N}$HR′

82 Put the following Z^- groups in order of decreasing stability:

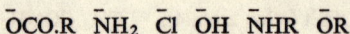

$$\bar{O}CO.R \quad \bar{N}H_2 \quad \bar{C}l \quad \bar{O}H \quad \bar{N}HR \quad \bar{O}R$$

A81

$$\underset{R}{\overset{O}{\underset{|}{H_2N-\overset{\parallel}{C}}}} + \bar{O}Et \text{ is more likely than } H_2\bar{N} + \underset{R}{\overset{O}{\underset{|}{\overset{\parallel}{C}-OEt}}}$$

$$\underset{R}{\overset{O}{\underset{|}{MeNH-\overset{\parallel}{C}}}} + \bar{O}CO.R \text{ is more likely than } Me\bar{N}H + \underset{R}{\overset{O}{\underset{|}{\overset{\parallel}{C}OCO.R}}}$$

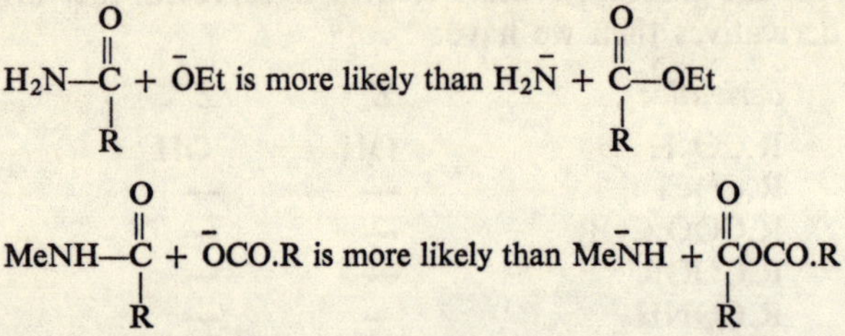

83 It is sometimes possible to produce the less likely product by modifying the reaction condition in some way. Remembering that the reactions are reversible can you suggest procedures for shifting the equilibrium in the desired direction?

A82

(most stable) Cl^-, $\bar{O}CO.R$, $\bar{O}R$, $\bar{O}H$, $\bar{N}HR$, $\bar{N}H_2$ (least stable).

---

84 These substitutions may be catalysed; sometimes with acids, sometimes with bases. As with aldehydes and ketones, basic catalysts increase the concentration and/or reactivity of the —— reagent.
We shall find that whereas acid catalysts can interact with the carbonyl *carbon/oxygen* atom they commonly affect the group Z also and facilitate its removal.

A83 We can apply the principles of the Law of Mass Action and use an excess of one of the reagents and/or remove a product as it is formed.

---

85 (i) Acyl halides

The acid chlorides are the most common of the acyl halides and our discussion will be confined to these.
The acyl chlorides are the most reactive of all the acyl derivatives because ——. Consequently they are described as the most active of the acylating agents. This statement gives the impression that the acid chloride attacks the other reagent but we prefer, here, to speak as though the other reagent attacks the acyl chloride. Neither description is more correct than the other.

The acyl chlorides are readily converted into all the other types of acid derivatives under consideration.

reaction	*general name of organic product*
R.COCl+H_2O ⟶ R.CO_2H+HCl	acid
R.COCl+EtOH ⟶ —— + ——	—
R.COCl+EtSH ⟶ —— + ——	thioester
R.COCl+R.COONa ⟶ —— + ——	—
R.COCl+NH_3 ⟶ —— + ——	—
R.COCl+R′NH_2 ⟶ —— + ——	—
R.COCl+NaN_3 ⟶ —— + NaCl	azide

A84 nucleophilic, oxygen.

---

86 These reactions proceed so readily that a catalyst is seldom needed and they follow the general mechanistic pathway which was outlined in Part 1:

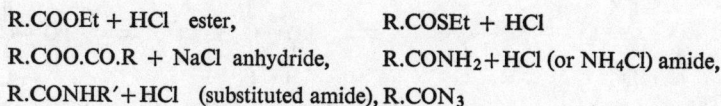

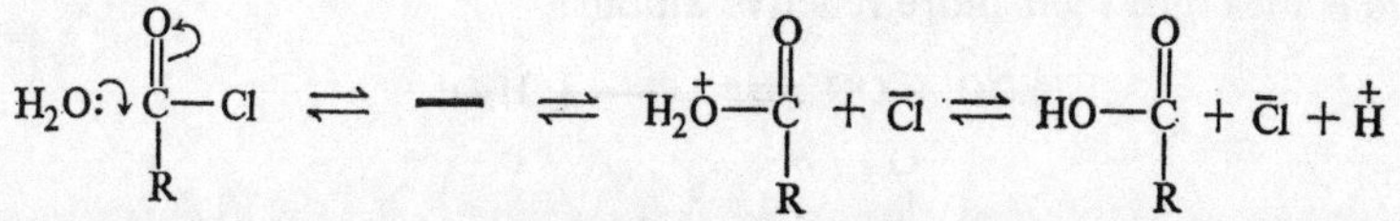

It is important not to get lost in symbols but to remember what reaction is being discussed. This sequence shows what happens when an —— reacts with ——. The organic product is a ——.

A85 Because the chloride ion is the most stable of the ions which are likely to be eliminated.

R.COOEt + HCl ester, R.COSEt + HCl

R.COO.CO.R + NaCl anhydride, R.$CONH_2$+HCl (or NH_4Cl) amide,

R.CONHR′+HCl (substituted amide), R.CON_3

---

87 When you formulate such a reaction you have to decide the exact nature of the nucleophilic reagent. With the above example it could be H_2O or $H\bar{O}$. The first is the weaker nucleophile but it is present in high concentration. The second is the stronger nucleophile but is present in low concentration particularly as the reaction mixture becomes acidic. Therefore, as shown in frame 86 the nucleophilic reagent is water.

What would you expect to be the active nucleophilic reagent in the reaction of an acyl halide with EtOH, R.COONa, NH_3, and NaN_3?

A86

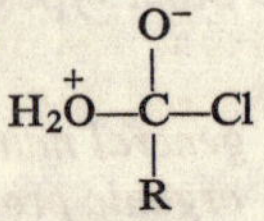

$$H_2\overset{+}{O}-\underset{R}{\overset{\overset{\displaystyle O^-}{|}}{C}}-Cl \quad \text{acid(acyl) chloride, water, carboxylic acid.}$$

----

88 Now formulate the reaction between propionyl chloride (Et.COCl) and hexanol and between acetyl chloride and sodium acetate. State briefly in words what these reactions are as in frame 86.

A87

EtOH R.CO$\bar{\text{O}}$ NH$_3$ $\bar{\text{N}}_3$ (azide ion)

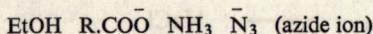

89 Many alcohols, phenols, and amines can be benzoylated by reaction with benzoyl chloride (the Schotten-Baumann reaction). Acid chlorides of aromatic acids are somewhat less reactive than the aliphatic compounds and this reaction is often effected in the presence of a base (alkali or tertiary amine). The function of the basic catalyst is not entirely clear but in the reaction with phenols it will convert these molecules into their more reactive anions.

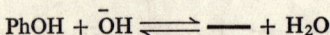

$$PhOH + \bar{O}H \rightleftharpoons \text{---} + H_2O$$

$$\text{---} + \underset{Ph}{\overset{\overset{\displaystyle O}{\|}}{C}}-Cl \rightleftharpoons \text{---} \rightleftharpoons \text{---} + Cl^-$$

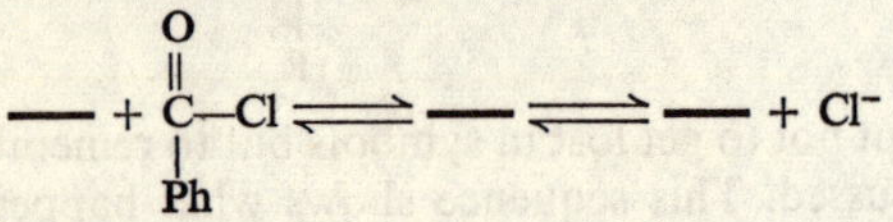

This is the benzoylation of —— by —— and furnishes phenyl benzoate.

A88 Hexanol is CH$_3$.CH$_2$.CH$_2$.CH$_2$.CH$_2$.CH$_2$OH; it will be represented as ROH

$$RO\overset{\curvearrowleft}{}\underset{\overset{|}{H}\ Et.}{\overset{\overset{\displaystyle O}{\|}}{C}}-Cl \rightleftharpoons RO-\underset{\overset{|}{H}\ Et}{\overset{\overset{\displaystyle O^-}{|}}{C}}\overset{\curvearrowleft}{}Cl \rightleftharpoons R\overset{+}{O}-\underset{\overset{|}{H}\ Et}{\overset{\overset{\displaystyle O}{\|}}{C}} + Cl^- \overset{-H^+}{\rightleftharpoons} RO-\underset{Et}{\overset{\overset{\displaystyle O}{\|}}{C}}$$

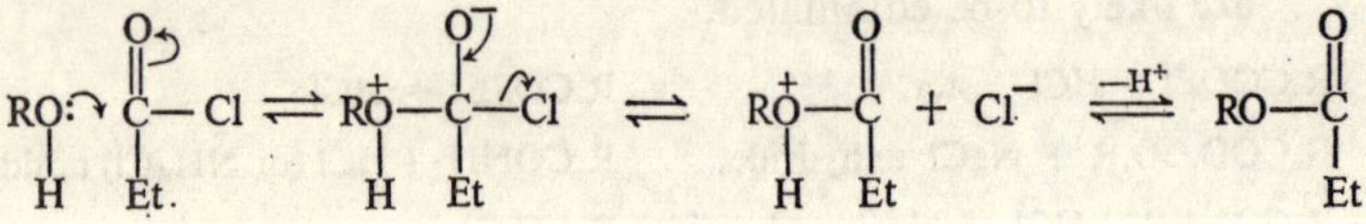

This is the preparation of the ester hexyl propionate from hexanol and propionyl chloride.

$$CH_3.CO\bar{O}\overset{\curvearrowleft}{}\underset{CH_3}{\overset{\overset{\displaystyle O}{\|}}{C}}-Cl \rightleftharpoons CH_3.COO-\underset{CH_3}{\overset{\overset{\displaystyle O^-}{|}}{C}}\overset{\curvearrowleft}{}Cl \rightleftharpoons CH_3.COO.\underset{CH_3}{\overset{\overset{\displaystyle O}{\|}}{C}} + Cl^-$$

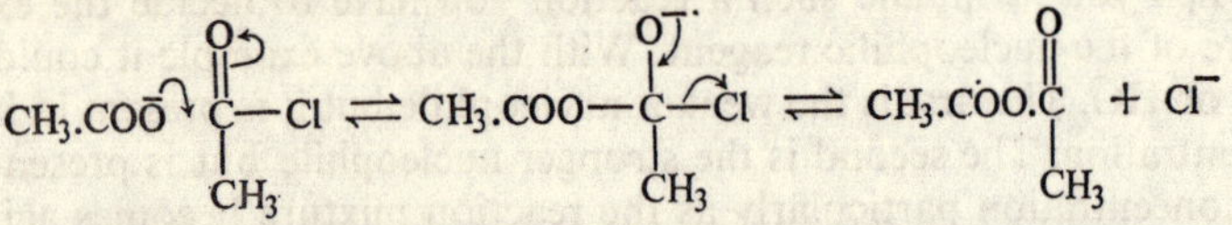

The preparation of an anhydride by the general reaction between an acid chloride and the sodium salt of the corresponding acid.

90 Another special catalysed reaction of acyl halides is the Friedel Crafts acylation reaction discussed later (see frames 95 and 96).

Continue with frame 91.

A89

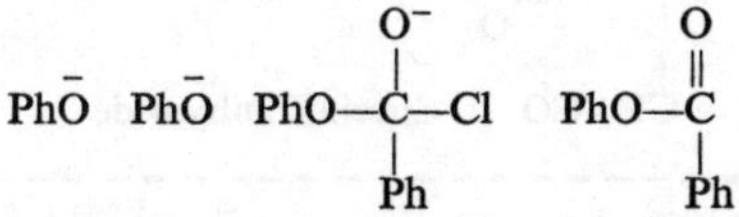

$$PhO^- \quad PhO^- \quad PhO-\overset{O^-}{\underset{Ph}{C}}-Cl \quad PhO-\overset{O}{\underset{Ph}{\overset{\|}{C}}} \quad \text{phenol, benzoyl chloride.}$$

---

91 (ii) Acid anhydrides

These are the next most reactive of the acyl derivatives. They react less vigorously than the acyl halides, but otherwise in a similar manner. The anhydrides contain two acyl groups but only one is retained in the product of interest, the other is lost as the carboxylic acid.

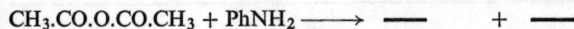

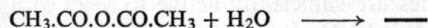

$$CH_3.CO.O.CO.CH_3 + EtOH \longrightarrow CH_3.COOEt + CH_3.CO_2H$$

$$CH_3.CO.O.CO.CH_3 + NH_3 \longrightarrow \underline{\quad\quad} + CH_3.CO_2H$$

$$CH_3.CO.O.CO.CH_3 + PhNH_2 \longrightarrow \underline{\quad\quad} + \underline{\quad\quad}$$

$$CH_3.CO.O.CO.CH_3 + H_2O \longrightarrow \underline{\quad\quad}$$

---

92 The reaction with ethanol occurs thus:

$$EtO-\overset{\overset{O}{\|}}{\underset{\underset{H}{|}\ \underset{CH_3}{|}}{C}}-O.CO.CH_3 \rightleftharpoons \underline{\quad} \rightleftharpoons \underline{\quad} + CH_3.CO_2^- \xrightarrow[\text{transfer}]{\text{proton}} \underline{\quad} + \underline{\quad}$$

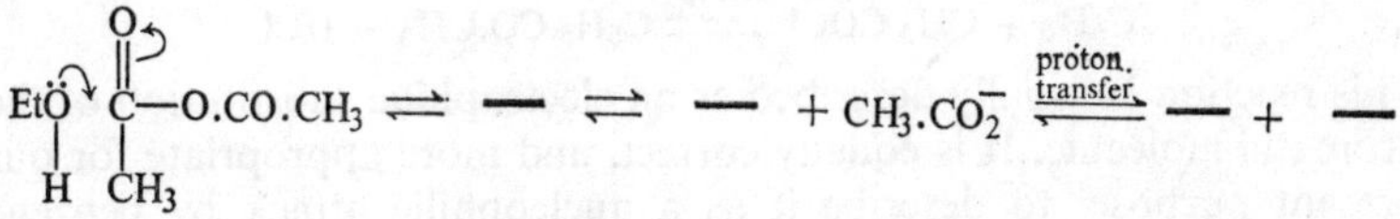

What is the name of the product?

A91 $CH_3.CONH_2$ $\quad$ $CH_3.CONHPh + CH_3.COOH$ $\quad$ $2CH_3.COOH$

---

93 Aniline ($PhNH_2$) and acetic anhydride form acetanilide. Formulate this reaction in a similar manner to that used in frame 92.

A92

$$EtO-\overset{O^-}{\underset{\underset{H}{|}\ \underset{CH_3}{|}}{\overset{+}{C}}}-O.CO.CH_3 \quad EtO-\overset{O}{\underset{\underset{H}{|}\ \underset{CH_3}{|}}{\overset{+}{C}}} \quad EtO-\overset{O}{\underset{\underset{CH_3}{|}}{\overset{\|}{C}}} \ +CH_3.CO_2H$$

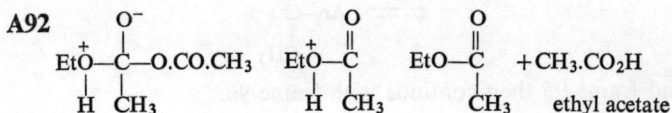

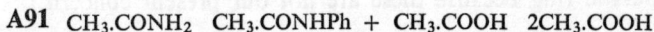

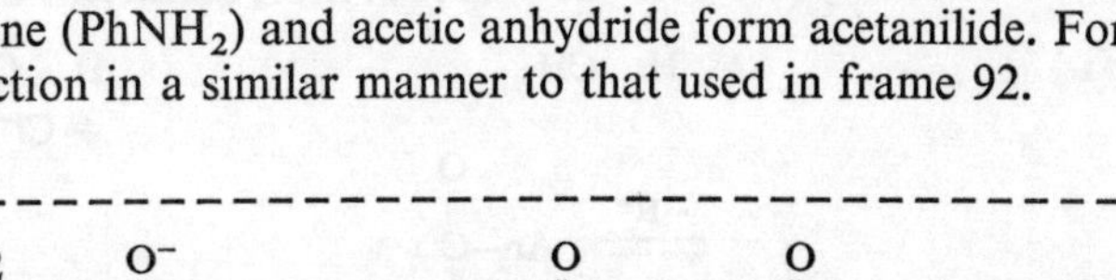

ethyl acetate

94 Can you suggest what would result from reaction between methanol and succinic anhydride (below)? Formulate the mechanism of the reaction. If you are unsure of the product follow the mechanism through first and see if you can derive the product.

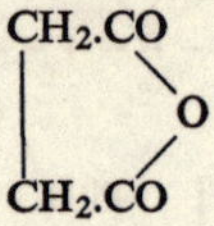

$$
\begin{array}{c}
CH_2.CO \\
\diagdown \\
O \\
\diagup \\
CH_2.CO \qquad \text{succinic anhydride}
\end{array}
$$

A93

$$
\underset{\underset{CH_3}{|}}{\overset{\overset{O}{\parallel}}{Ph\overset{+}{N}_{H_2}-C}}-OCO.CH_3 \ \rightleftharpoons \ \underset{\underset{CH_3}{|}}{\overset{\overset{O^-}{|}}{Ph\overset{+}{N}_{H_2}-C}}-OCO.CH_3 \ \rightleftharpoons \ \underset{\underset{CH_3}{|}}{\overset{\overset{O}{\parallel}}{Ph\overset{+}{N}_{H_2}-C}} + CH_3.COO^-
$$

$$
\rightleftharpoons \ \underset{\underset{CH_3}{|}}{\overset{\overset{O}{\parallel}}{PhNHC}} + CH_3.COOH
$$

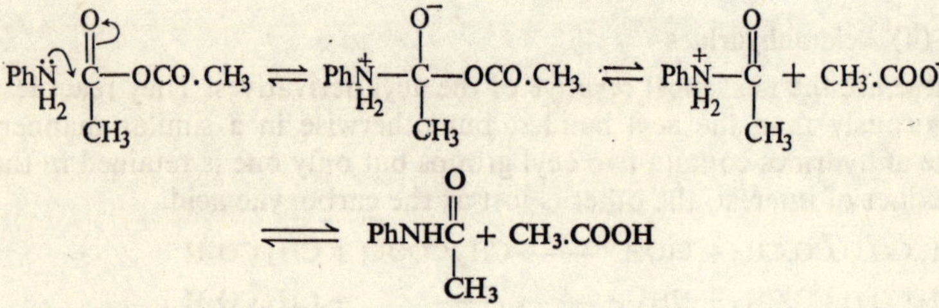

95 Acid chlorides and anhydrides are sufficiently active to react with very weak nucleophiles. In the **Friedel Crafts reaction** aromatic molecules such as benzene react with acid chlorides or anhydrides in the presence of catalysts such as aluminium trichloride,

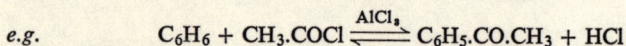

e.g. $\qquad C_6H_6 + CH_3.COCl \overset{AlCl_3}{\rightleftharpoons} C_6H_5.CO.CH_3 + HCl$

This reaction is usually described as an electrophilic substitution of the aromatic molecule. It is equally correct, and more appropriate for our present purpose, to describe it as a nucleophilic attack by benzene on the acyl halide or anhydride. The catalyst acts on the chlorine atom (as well as on the oxygen atom) thereby increasing further the activity of the carbonyl function. (We shall ignore the electron movements in the aromatic ring because these are not our present concern.)

$$
\underset{\underset{H \quad CH_3}{|\qquad|}}{\overset{\overset{O}{\parallel}}{Ar\ \ C}}-Cl\ldots\ldots AlCl_3 \rightleftharpoons \underset{\underset{H \quad CH_3}{|\qquad|}}{\overset{\overset{O^-}{|}}{Ar-\overset{+}{C}l}}-Cl\ldots\ldots AlCl_3 \rightleftharpoons \underset{\underset{H \quad CH_3}{|\qquad|}}{\overset{\overset{O}{\parallel}}{Ar-\overset{+}{C}}}
$$

$$
+ \ Cl^- + AlCl_3
$$

$$
\overset{-H^+}{\rightleftharpoons} \ \underset{\underset{CH_3}{|}}{\overset{\overset{O}{\parallel}}{Ar-C}}
$$

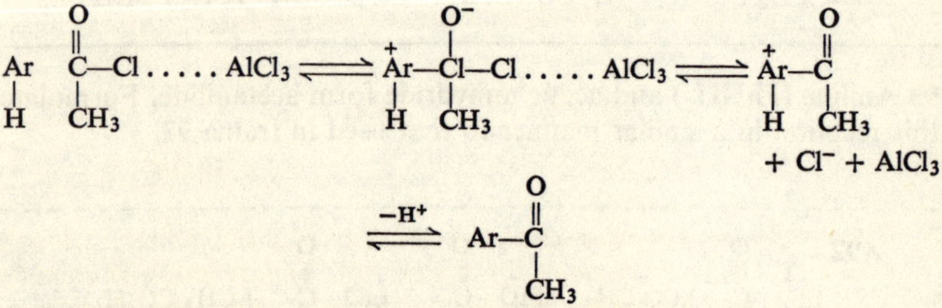

Reread frame 95 then continue with frame 96.

A94 $MeO.CO.CH_2.CH_2.COOH$

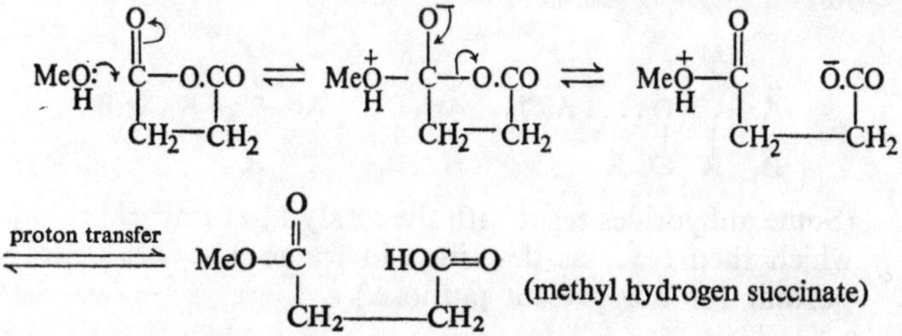

(methyl hydrogen succinate)

96 We can bring the Friedel Crafts acylation reaction into our general picture of carbonyl reactions by considering it as attack by a weak —— (*viz.* the aromatic molecule) on the two most active acyl derivatives (*viz.* —— and ——) which are further activated by the catalyst (——). The product is an aromatic *hydrocarbon/ketone/acid.*

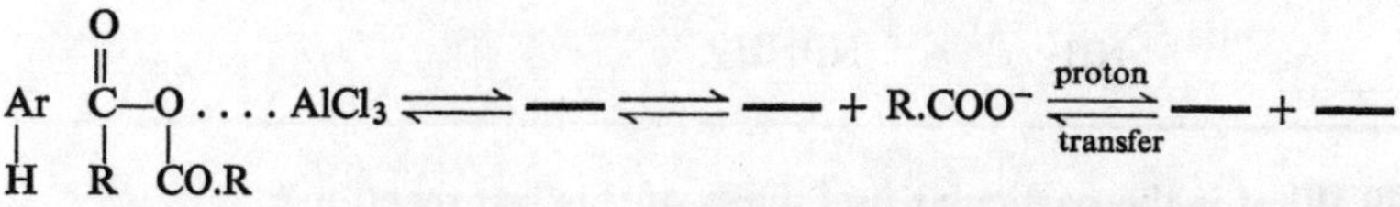

97 (iii) Esters

By the appropriate substitution reaction, esters (R.COOR′) can be converted into carboxylic acids ($R.CO_2H$), amides ($R.CONH_2$), hydrazides ($R.CONHNH_2$), and hydroxamic acids (RCONHOH). The extremely important interconversion of acids and esters will be discussed later; the other reactions will be considered more briefly.

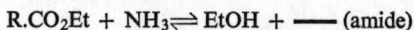

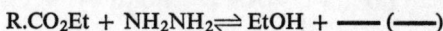

$$R.CO_2Et + NH_3 \rightleftharpoons EtOH + \text{——} \text{ (amide)}$$
$$R.CO_2Et + NH_2NH_2 \rightleftharpoons EtOH + \text{——} \text{ (——)}$$
$$R.CO_2Et + NH_2OH \xrightarrow{KOH} EtOH + R.C(=O)NHOH \quad \text{(——)}$$

The last reaction is base-catalysed and therefore the actual nucleophilic species is likely to be —— (*cf.* frame 40). This reaction is sometimes used as a test for esters as the hydroxamic acids give characteristic colours with ferric chloride solution.

A96 nucleophile, acyl halides (chlorides) and anhydrides, aluminium trichloride, ketone

$$\underset{\underset{H\ \ R\ \ CO.R}{|\ \ \ |}}{\overset{O^-}{\underset{|}{Ar-\overset{+}{C}-O}}}\dots AlCl_3 \quad \underset{\underset{H\ \ R}{|\ \ \ |}}{\overset{O}{\overset{||}{Ar-\overset{+}{C}}}} \quad \underset{\underset{R}{|}}{\overset{O}{\overset{||}{Ar-C}}} + R.CO_2H$$

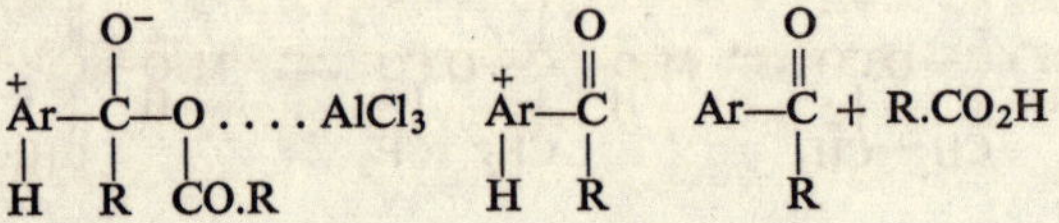

(Some anhydrides react with the catalyst to form acid chlorides which then react as described in frame 95. This is not important for our present purpose.)

---

98 Formulate the substitution reaction between methyl benzoate (Ph.COOMe) and (i) ammonia, (ii) NH_2OH and KOH.

A97

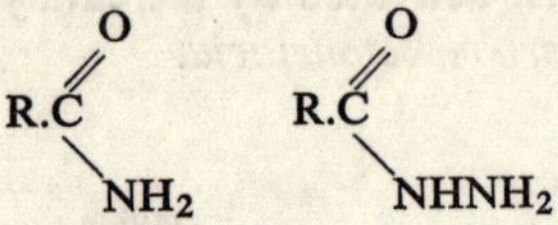

$$\underset{NH_2}{\overset{O}{\overset{||}{R.C}}} \qquad \underset{NHNH_2}{\overset{O}{\overset{||}{R.C}}} \qquad \text{(hydrazide), (hydroxamic acid), } \overline{N}HOH$$

---

99 What is the particular usefulness of this last reaction?

A98

$$H_3\overset{+}{N}-\underset{Ph}{\overset{O}{\overset{||}{C}}}-OMe \rightleftharpoons H_3\overset{+}{N}-\underset{Ph}{\overset{O^-}{\overset{|}{C}}}-OMe \rightleftharpoons H_3\overset{+}{N}-\underset{Ph}{\overset{O}{\overset{||}{C}}} + \overline{O}Me \rightleftharpoons H_2N-\underset{Ph}{\overset{O}{\overset{||}{C}}} + MeOH$$

$$HO\overset{}{N}H-\underset{Ph}{\overset{O}{\overset{||}{C}}}-OMe \rightleftharpoons HONH-\underset{Ph}{\overset{O^-}{\overset{|}{C}}}-OMe \rightleftharpoons HONH-\underset{Ph}{\overset{O}{\overset{||}{C}}} + \overline{O}Me$$

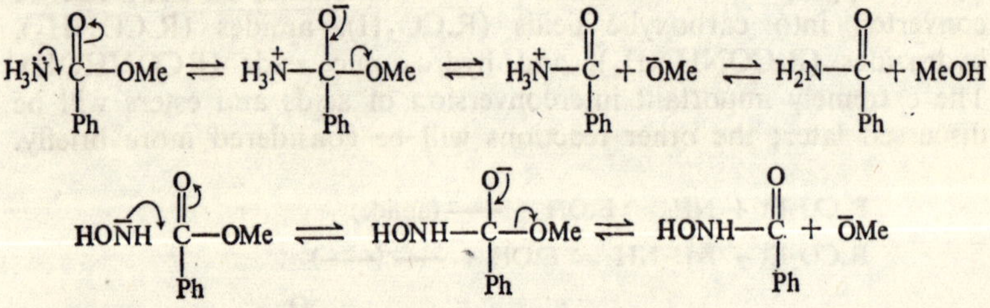

100 Can you suggest a reason why acid chlorides and anhydrides cannot be prepared directly from esters?

A99 It is used as a test for esters. The hydroxamic acid is recognised by its colour when mixed with ferric chloride solution.

101 Esters also enter into several important reactions in which the attacking nucleophile is a carbanion (see frames 50–53). The self-condensation of esters falls into this category. One ester molecule provides the carbanion through interaction with basic catalyst and a second ester molecule acts as a carbonyl compound.

For example, ethyl acetate reacts with sodium to give ethyl acetoacetate.

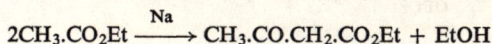

$$2CH_3.CO_2Et \xrightarrow{\text{Na}} CH_3.CO.CH_2.CO_2Et + EtOH$$

This reaction is discussed in detail in '*Programmes in Organic Chemistry*' Volume 2. It is called the **Claisen ester condensation.**
Continue with frame 102.

A100 Reaction between ester and chloride ion might give an inter-

mediate of type

$$Cl-\underset{\underset{R}{|}}{\overset{\overset{O^-}{|}}{C}}-OEt$$

but this would eliminate the more stable Cl^- and regenerate the ester, rather than acid halide and $\bar{O}Et$.

The same argument holds for reaction with a sodium carboxylate. The intermediate would eliminate carboxylate anion ($R.COO^-$) rather than ethoxide ion.

---

102 The basic catalyst is ethoxide ion produced from sodium and a trace of ethanol.

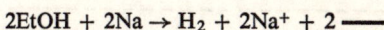

$$2EtOH + 2Na \rightarrow H_2 + 2Na^+ + 2 \text{---}$$

This reacts with ethyl acetate to furnish the c——n which is a resonance-stabilised system.

$$CH_3-\underset{\underset{OEt}{|}}{C}=O + \bar{O}Et \rightleftharpoons \text{---} \leftrightarrow \text{---}$$

---

103 The rest of the reaction falls into the general pattern we have now established.

$$EtO_2C.\overset{\frown}{\bar{C}H_2} \overset{\overset{O}{\|}}{\underset{\underset{CH_3}{|}}{C}}-OEt \rightleftharpoons \text{---} \rightleftharpoons \text{---} + \bar{O}Et$$

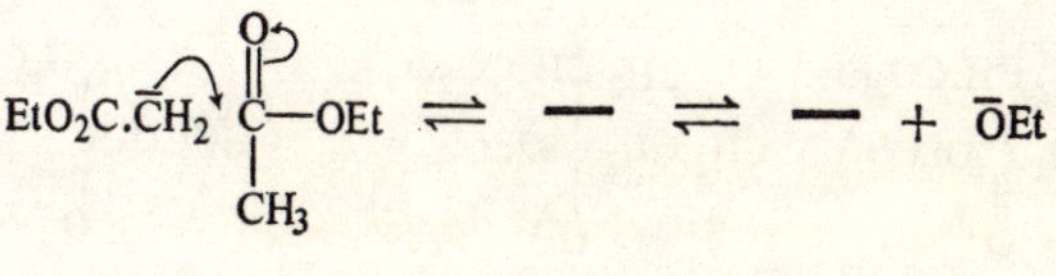

A102 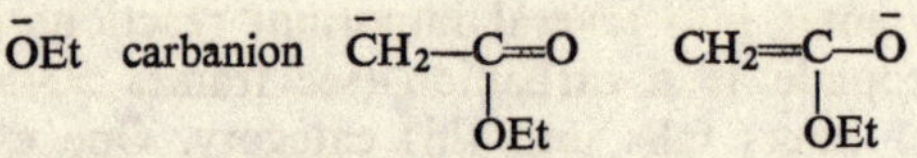

104 Esters of dibasic acids can react in an intramolecular manner:

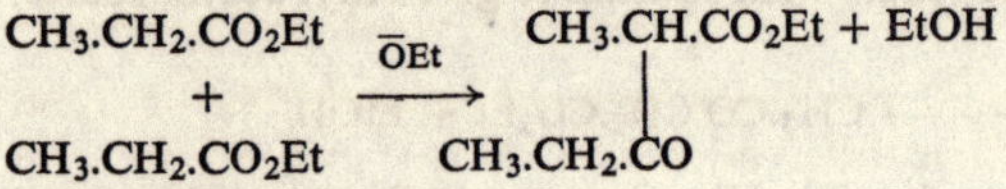

ethyl propionate (2 mol)

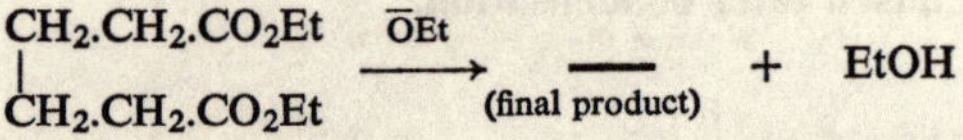

ethyl adipate (1 mol)

Formulate the monovalent carbanion you would expect to get from ethyl adipate.

A103 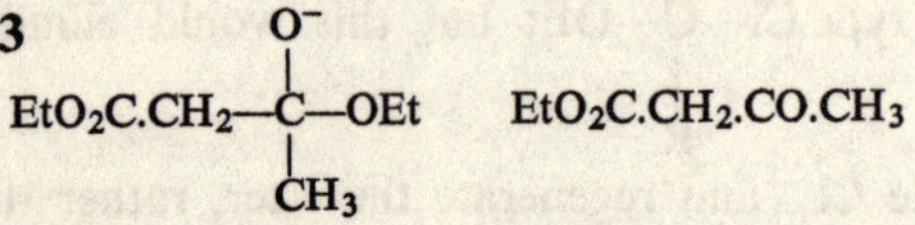

105 Now formulate the whole reaction

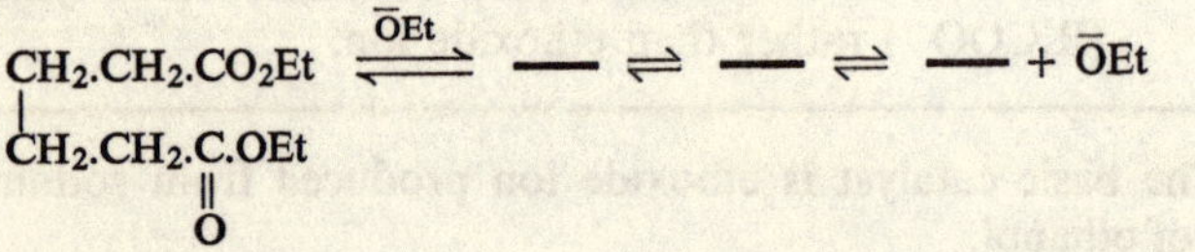

A104 

106 Self condensation is only possible when the ester can form a carbanion. Can you suggest any ester which cannot form a carbanion and therefore will not condense with itself? If you have difficulty here consult frame 60.

A105

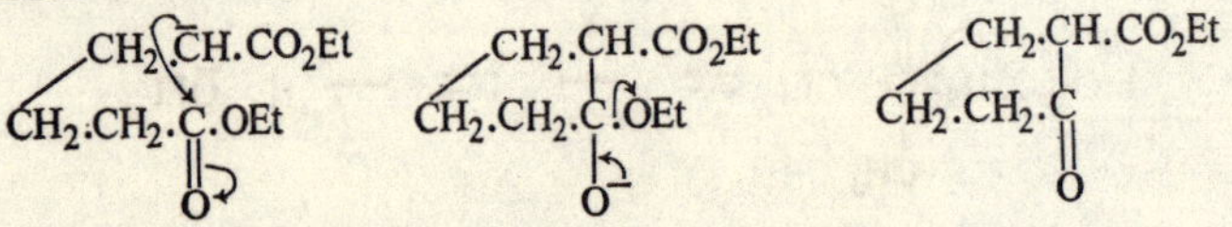

107 Name the three substances referred to in the above answer.

A106 Esters which do not have any hydrogen atoms on the α-carbon atom (*e.g.* Ph.COOEt) or do not have an α-carbon atom (*e.g.* HCOOEt or EtOOC.COOEt) cannot form a carbanion.

----

108 These esters however, can react with carbanions produced from other molecules, including other esters,

$$e.g.\ \text{Ph.CO}_2\text{Et} + \text{CH}_3\text{.CO}_2\text{Et} \underset{}{\overset{\overline{O}Et}{\rightleftharpoons}} \text{——} + \text{EtOH}$$
$$\text{(final product)}$$

A107 ethyl benzoate, ethyl formate, and ethyl oxalate.

----

109 In more detail this reaction proceeds as follows:

$$\text{EtO}_2\text{C.CH}_3 \overset{\overline{O}Et}{\rightleftharpoons} \text{——} + \underset{\underset{\text{Ph}}{|}}{\overset{\overset{\text{O}}{\|}}{\text{C}}}\text{—OEt} \rightleftharpoons \text{——} \rightleftharpoons \text{——} + \overline{O}Et$$

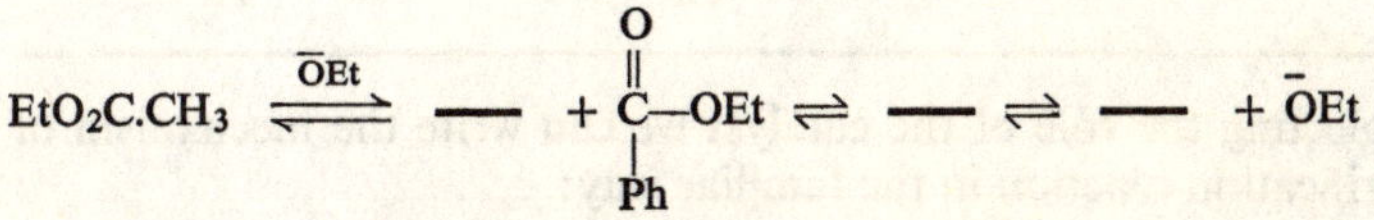

A108   Ph.CO.CH$_2$.CO$_2$Et

----

110 What other product might be formed in this reaction between ethyl acetate and ethyl benzoate?

A109

$$\text{EtO}_2\text{C.}\overline{C}\text{H}_2 \qquad \text{EtO}_2\text{C.CH}_2\text{—}\underset{\underset{\text{Ph}}{|}}{\overset{\overset{O^-}{|}}{C}}\text{—OEt} \qquad \text{EtO}_2\text{C.CH}_2\text{.CO.Ph}$$

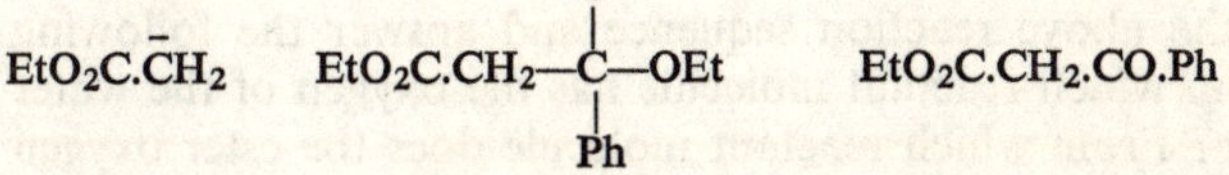

111 Remember, finally, that the carbanion may be obtained from molecules other than esters (see frames 50–53).

$$e.g.\ \text{CH}_3\text{.CO.CH}_3 \overset{\text{base}}{\underset{\text{carbanion}}{\rightleftharpoons}} \text{——} + \underset{\underset{\text{Ph}}{|}}{\overset{\overset{\text{O}}{\|}}{\text{C}}}\text{—OEt} \rightleftharpoons \text{——} \rightleftharpoons \text{——} + \overline{O}Et$$

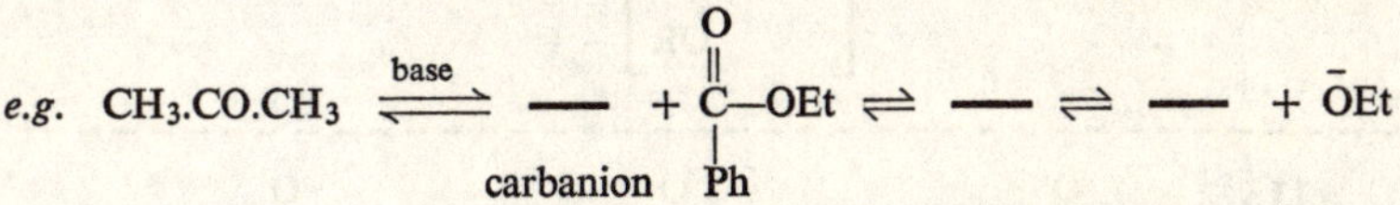

A110   CH$_3$.CO.CH$_2$.CO$_2$Et (by self condensation of ethyl acetate).

----

112 We return now to the very important reactions of ester-formation and ester-hydrolysis. Hydrolysis is catalysed by acids or bases, esterification only by acids.
Alcohols are converted into esters by acylation and this can be effected by an acid chloride, or anhydride, or by the acid itself. Only the last of these requires a catalyst.

ROH + R′.COCl → —— + ——
ROH + R′.CO.O.COR′ → —— + ——
ROH + R′.COOH $\overset{H^+}{\rightleftharpoons}$ —— + ——

We are now concerned only with the last of these which, it should be noted, is a(n) *reversible/irreversible* reaction.

A111

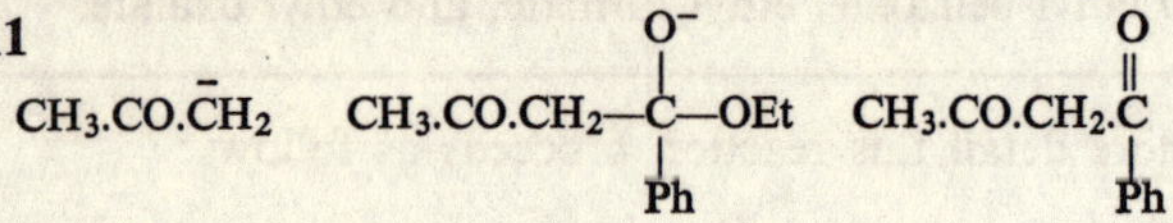

---

113 Neglecting the role of the catalyst we can write the mechanism of the esterification reaction in the familiar way:

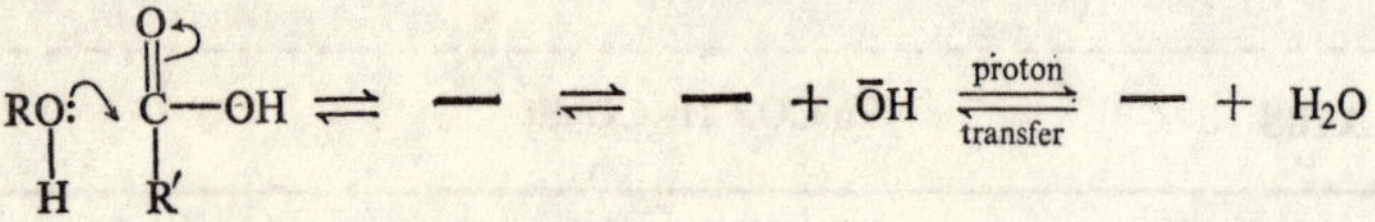

A112 R′.COOR + HCl   R′.COOR + R′.COOH   R′.COOR + H_2O
reversible.

---

114 Examine the above reaction sequence and answer the following questions. From which reactant molecule has the oxygen of the water molecule come? From which reactant molecule does the ester oxygen marked * arise

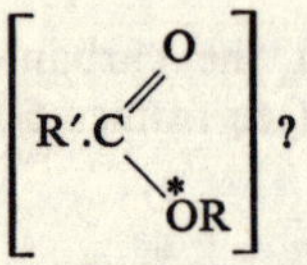

A113

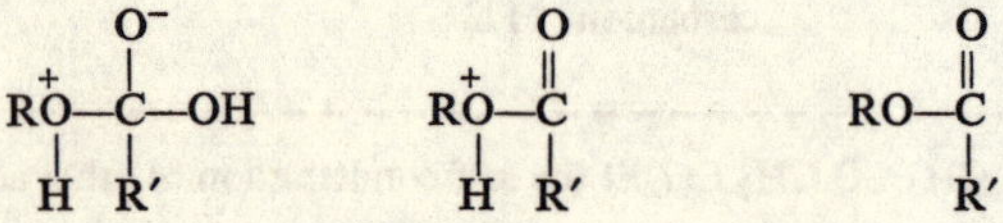

115 In the absence of catalyst this esterification reaction is generally impracticably slow. Reaction involves attack by the nucleophilic reagent (——) on the carbonyl carbon of the carboxyl group. Now we have to discover the role of the catalyst. Would protonation of the alcohol increase or decrease its nucleophilic character?

- -

A114 from the acid molecule, from the alcohol molecule.

- -

116 The catalyst must influence, therefore, the carboxylic acid but there are two oxygen atoms which could be protonated. Formulate the two oxonium ions one of which can also be written as a carbonium ion.

- -

A115 ROH (or alcohol),
protonation would decrease the nucleophilic reactivity of the alcohol.

- -

117 Both of these will be more reactive than the carboxylic acid which is a very weak carbonyl compound. It is now believed that esterification

proceeds through the intermediate $R-\overset{\overset{\displaystyle O}{\|}}{C}-\overset{+}{O}H_2$ and the sequence in frame 113 can be written more correctly:

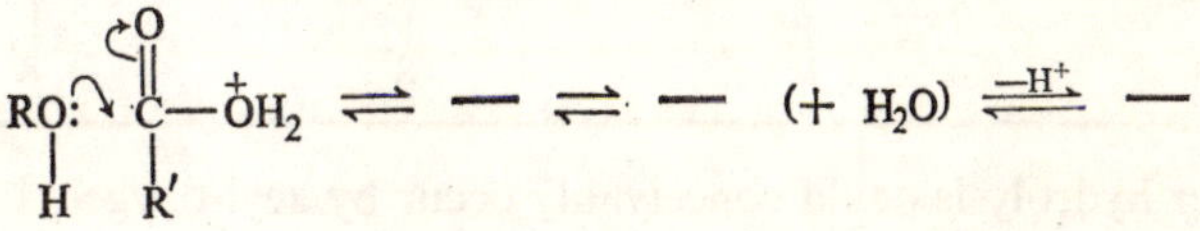

$$RO\overset{\curvearrowright}{:} \underset{\underset{H}{|}}{\overset{\overset{\displaystyle\curvearrowleft O}{\|}}{C}}\underset{R'}{\overset{+}{-O}H_2} \rightleftharpoons \; - \; \rightleftharpoons \cdot \; - \quad (+ \; H_2O) \xrightarrow{\underset{}{-H^+}} \; -$$

- -

A116

$$R-\overset{\overset{+}{O}H}{\underset{\|}{C}}-OH \;\; \leftrightarrow \;\; R-\overset{OH}{\underset{|}{\overset{+}{C}}}-OH \qquad R-\overset{\overset{\displaystyle O}{\|}}{C}-\overset{+}{O}H_2$$
$$\text{(carbonium ion)}$$

- -

118 Acid-catalysed hydrolysis occurs by the reversal of the process shown in frame 117 and involves attack by water (as a(n) *electrophilic/nucleophilic* reagent) on the ester protonated on its alkoxy oxygen.

$$HO\overset{\curvearrowright}{:} \underset{\underset{H}{|} \;\; \underset{R'}{|} \;\; \underset{H}{|}}{\overset{\overset{\displaystyle O\curvearrowright}{\|}}{C}}-\overset{+}{O}R \rightleftharpoons \; - \; \rightleftharpoons \; - \quad (+ROH) \xrightarrow{\underset{}{-H^+}} \; -$$

- -

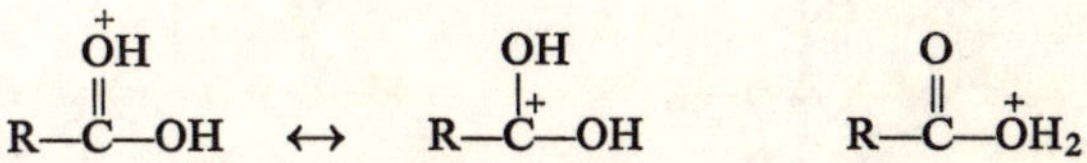

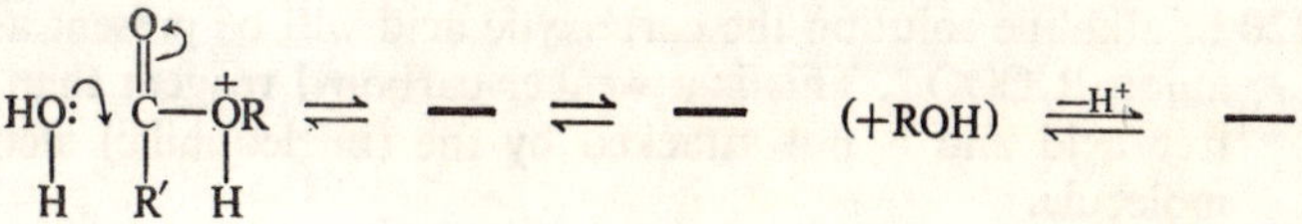

A117

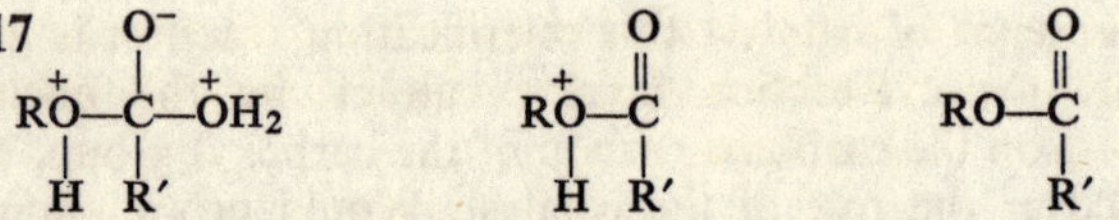

119 Base-catalysed hydrolysis involves attack by hydroxyl ion on the ester molecule and given that the sequence ends with an irreversible proton transfer you should be able to work it out. Try it.

A118 nucleophilic

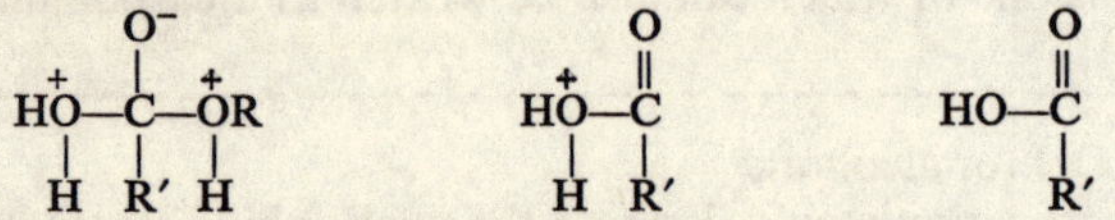

120 Can you suggest why esterification does not occur in alkaline solution?

A119

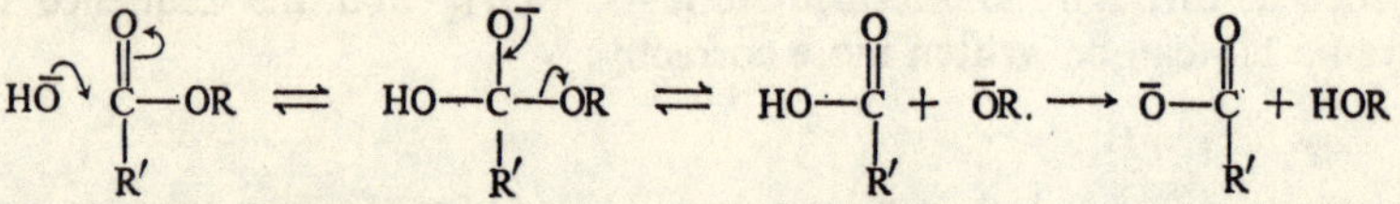

121 Ester hydrolysis could conceivably occur by acyl-oxygen fission or by alkyl-oxygen fission.

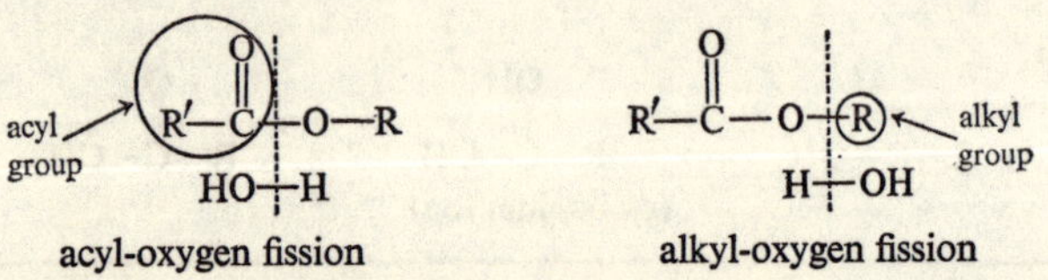

Consult frame 119 to see which is involved in ester hydrolysis.

A120 In alkaline solution the carboxylic acid will be present as its anion R.COO⁻. This is a weaker carbonyl reagent than the free acid and is not attacked by the (nucleophilic) alcohol molecule.

122 These ideas can be extended to include the direct conversion of one ester to another by reaction with an alcohol in the presence of a catalyst which may be acid or base,

e.g. $$R.COOR' + R''OH \overset{\text{catalyst}}{\rightleftharpoons} R.COOR'' + R'OH$$

Write the mechanism for the acid-catalysed *hydrolysis* of the ester R.COOR′.

A121 Reaction occurs by acyl-oxygen fission (alkyl-oxygen fission is known but is uncommon and will not be discussed here).

---

123 Now re-write this replacing the nucleophile water by the alcohol R″OH.

A122

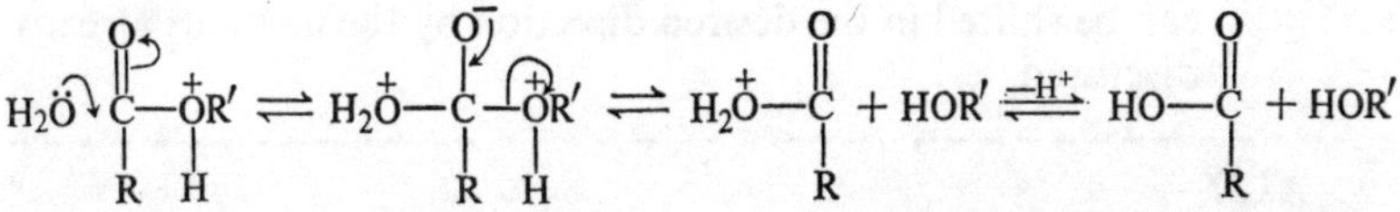

---

124 The starting material is an ester of the alcohol ——, the product is an ester of the alcohol ——. In fact the result will be an equilibrium mixture of two esters and two alcohols. Can you suggest any methods by which you could get a good yield of the ester R.COOR″?

A123

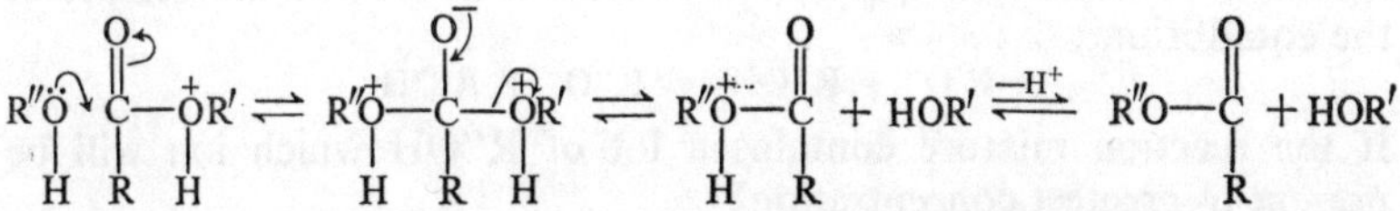

---

125 The base-catalysed interaction of alcohol (R″OH) and ester (R.COOR′) is usually effected by a little sodium. What do you think is the actual basic catalyst?

A124 R′OH R″OH
Use an excess of an alcohol R″OH and or remove one of the products of the reaction (R.COOR″ orR′/OH) as it is formed. If one of these is the most volatile component in the equilibrium mixture it could be distilled out.

126 Write the mechanism for the reaction between ester (R.COOR′) and hydroxyl ion. (Consult frame 119 if you have any difficulty.)

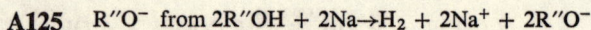

A125   R″O⁻ from 2R″OH + 2Na→H₂ + 2Na⁺ + 2R″O⁻

---

127 Rewrite this sequence (excluding the final irreversible step) replacing HO⁻ by R″O⁻.

A126

$$HO \overset{\displaystyle O}{\underset{\underset{R}{|}}{\overset{\|}{C}}} \!-\! OR' \rightleftharpoons HO \!-\! \overset{\displaystyle O^{-}}{\underset{\underset{R}{|}}{C}} \!-\! OR' \rightleftharpoons HO \!-\! \overset{\displaystyle O}{\underset{\underset{R}{|}}{\overset{\|}{C}}} + \; \bar{O}R' \longrightarrow \bar{O} \!-\! \overset{\displaystyle O}{\underset{\underset{R}{|}}{\overset{\|}{C}}} + \; HOR'$$

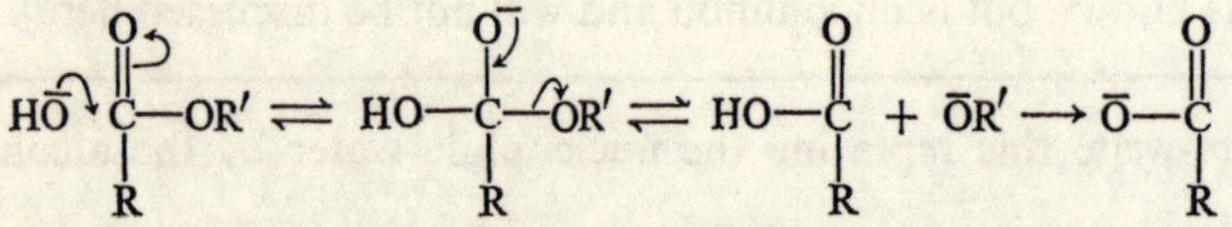

128 The product will be an equilibrium mixture of two alcohols (——— and ———) and two esters (——— and ———) and the equilibrium can be shifted in the desired direction by the methods already discussed.

A127

$$R''\bar{O} \overset{\displaystyle O}{\underset{\underset{R}{|}}{\overset{\|}{C}}} \!-\! OR' \rightleftharpoons R'O \!-\! \overset{\displaystyle O^{-}}{\underset{\underset{R}{|}}{C}} \!-\! OR' \rightleftharpoons R''O \!-\! \overset{\displaystyle O}{\underset{\underset{R}{|}}{\overset{\|}{C}}} + \; \bar{O}R'$$

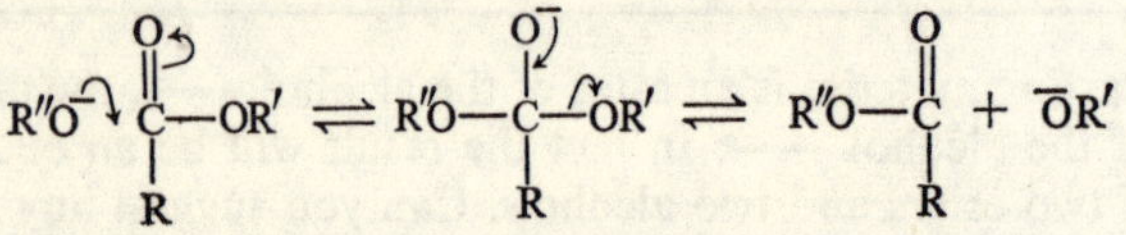

129 Go back to frame A127 for a moment. To get large amounts of R.COOR″ we need large amounts of R″O⁻. Where does this come from if only a catalytic quantity of sodium is used? The answer lies in the equilibrium:

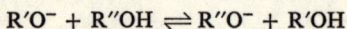

$$R'O^- + R''OH \rightleftharpoons R''O^- + R'OH$$

If the reaction mixture contains a lot of R″OH which ion will be present in greatest concentration?

A128   R′OH and R″OH   R.COOR′ and R.COOR″

---

130 Esters of the trihydric alcohol glycerol (fats) can be converted directly into their methyl esters by reaction with excess of methanol. What additional substances are needed before this reaction proceeds at a useful rate?

A129 R″O⁻ since the excess of R″OH displaces the equilibrium to the right.

131 (iv) Amides

The amides ($R.CONH_2$) contain a very weak carbonyl function and there is no satisfactory way of converting them directly to esters, anhydrides, or acid chlorides. They can be hydrolysed to the corresponding acids in the presence of acid or alkali.

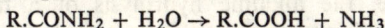

$$R.CONH_2 + H_2O \rightarrow R.COOH + NH_3$$

Alkaline hydrolysis follows the now familiar route and ends with a proton-transfer reaction. You should have no difficulty in completing the reaction sequence with HO^- and $R.CONH_2$. Try it.

A130 A basic catalyst, produced by addition of a little sodium, or an acidic catalyst, such as hydrogen chloride or sulphuric acid, is needed.

---

132 Acid hydrolysis is a reaction between water and the amide protonated on the nitrogen atom. Try this with the substituted amide R.CONHR′.

A131

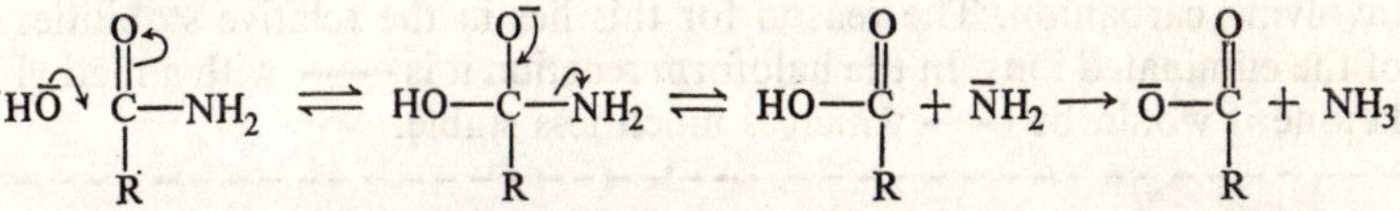

---

133 (v) Some further reactions of ketones

It is now apparent that much of the chemistry of aldehydes, ketones, and carboxylic acids and their derivatives can be understood in terms of closely related addition and substitution reactions which we have expressed in the terms:

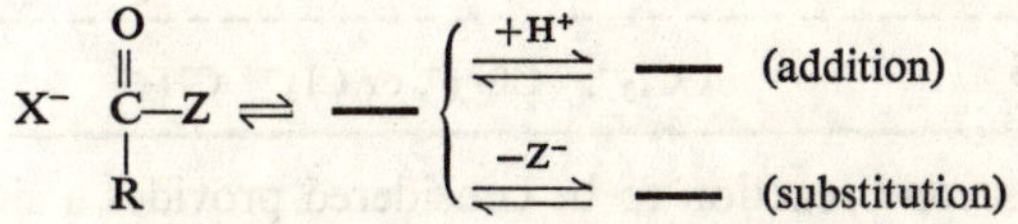

Aldehydes and ketones react by the *addition/substitution* route and the acyl compounds by the *addition/substitution* route.

A132

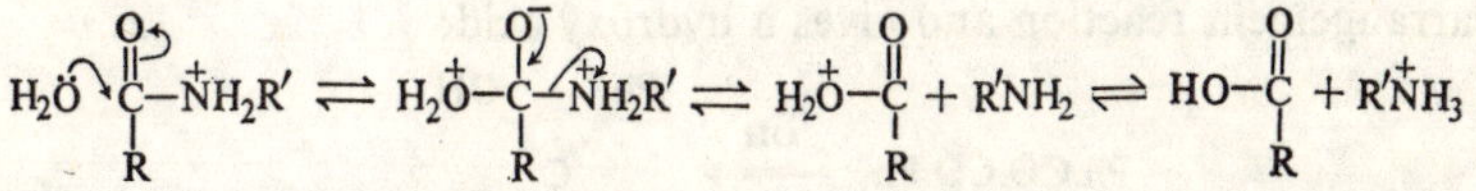

---

134 Finally, we shall examine two reactions of aldehydes and ketones which follow the substitution pathway or a simple extension of it. The reader is probably familiar with the reaction by which chloroform

($CHCl_3$), bromoform ($CHBr_3$), and iodoform (CHI_3) can be obtained from methyl ketones and a few related compounds by reaction with the appropriate halogen and alkali. This is often called the haloform reaction.

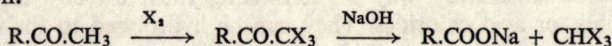

$$R.CO.CH_3 \xrightarrow{X_2} R.CO.CX_3 \xrightarrow{NaOH} R.COONa + CHX_3$$

The second stage, the alkaline hydrolysis of a trihalogenomethyl ketone, is of special interest for it is a typical *substitution* reaction.

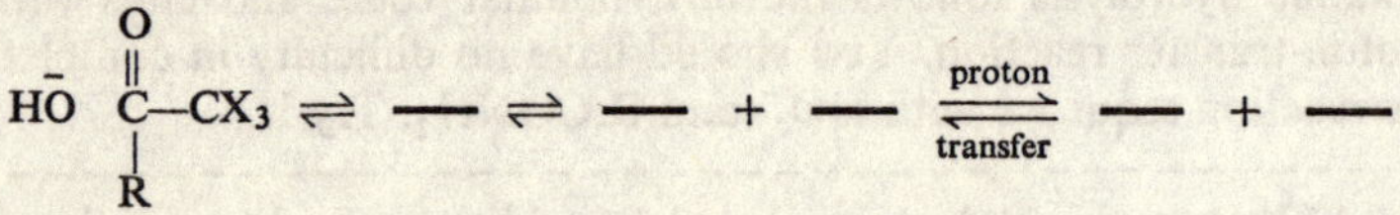

A133

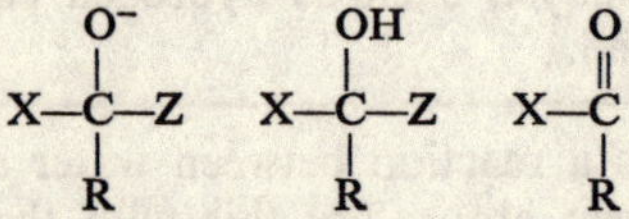

addition substitution

---

135 Methyl ketones are not hydrolysed like these trihalogenomethyl ketones though they will react with alkali in a condensation reaction involving carbanion. The reason for this lies in the relative stabilities of the eliminated ions. In the haloform reaction it is ——, with a methyl ketone it would be —— which is much less stable.

A134

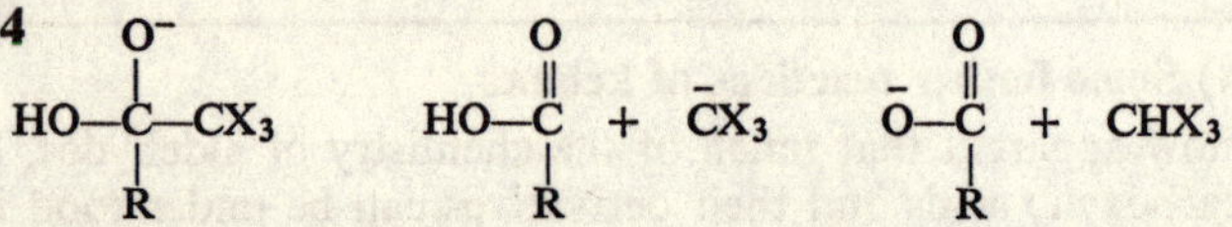

---

136 Can you suggest what causes the increased stability of CX_3^-?

A135 CCl_3^-, CBr_3^-, or CI_3^- CH_3^-

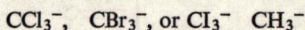

137 The second reaction to be considered provides a simple modification of the elimination of Z^-. The group Z is not eliminated but migrates from the carbonyl carbon atom to an adjacent electron-deficient centre. This is, in fact, a reaction of α-diketones.
When an α-diketone like benzil is treated with alkali it undergoes a rearrangement reaction and gives a hydroxy acid:

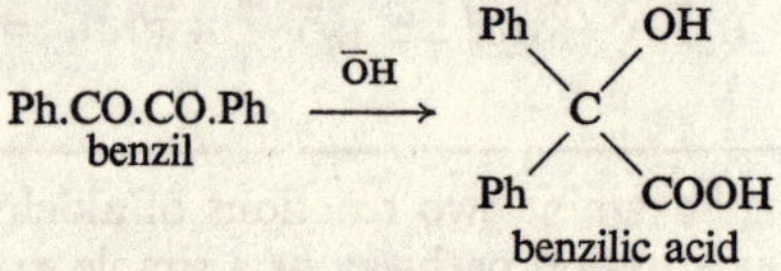

benzilic acid

In benzil the two phenyl groups are attached to the *same/different* carbon atom; in benzilic acid they are on the *same/different* carbon atom. Can you suggest a structure for the first stage of the sequence of reactions?

A136 The strong inductive effect from three C—X bonds.

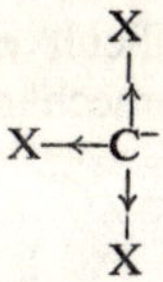

Apparently three of these are required before hydrolysis occurs.

---

138 Reaction starts by attack of HO^- on a carbonyl carbon atom and this is followed by migration (a special sort of elimination) of a phenyl group to the adjacent group thus:

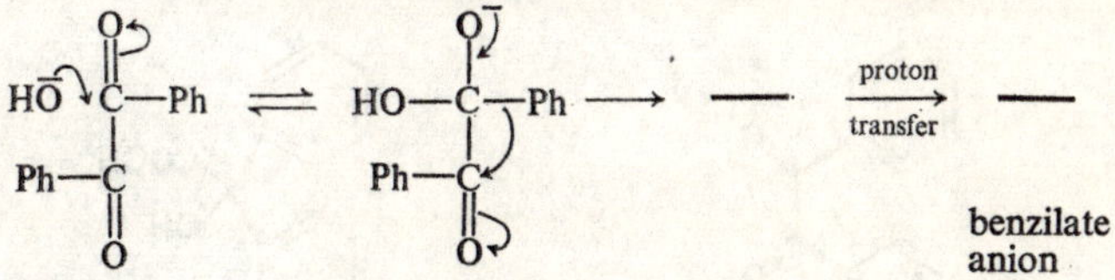

A137

different, the same, 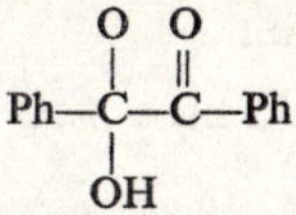

---

139 This reaction is called the **benzilic acid rearrangement.**
We know now enough chemistry to formulate a synthesis of benzilic acid from benzaldehyde in three stages:

PhCHO (benzaldehyde) ⇌ Ph.CH(OH).CO.Ph (benzoin) —oxidation→ Ph.CO.CO.Ph (benzil) → $Ph_2CH(OH).CO_2H$ (benzilic acid)

What reagent is required for (i) benzoin condensation (frame 22) and (ii) benzilic acid rearrangement?

A138

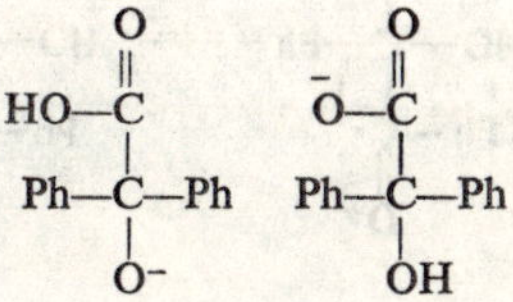

---

140 Elaborate the mechanisms of these two reactions.

A139 $\bar{C}N$ (from sodium cyanide), $\bar{O}H$ (from sodium hydroxide).

---

141 Here is a slightly more difficult example of a benzilic acid re-arrangement. Try to set out the mechanism for the conversion:

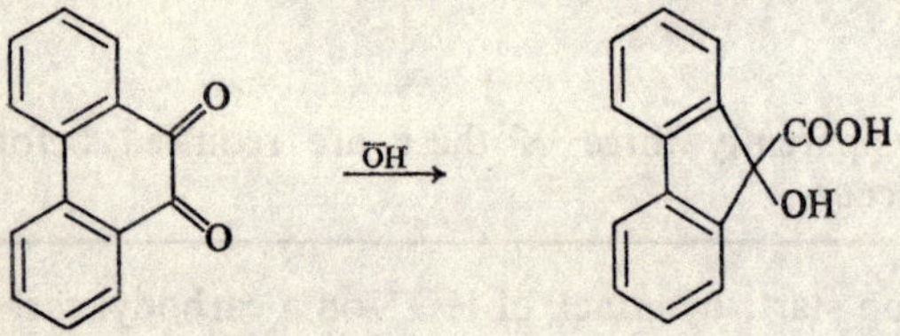

If you lack confidence notice that benzil and benzilic acid as written below are very similar to these tricyclic structures.

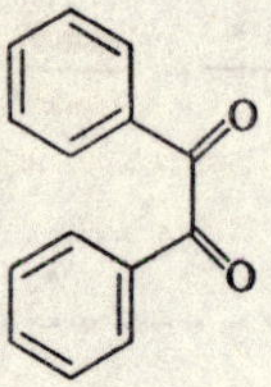

benzil

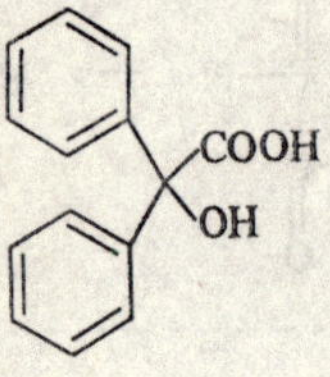

benzilic acid

A140

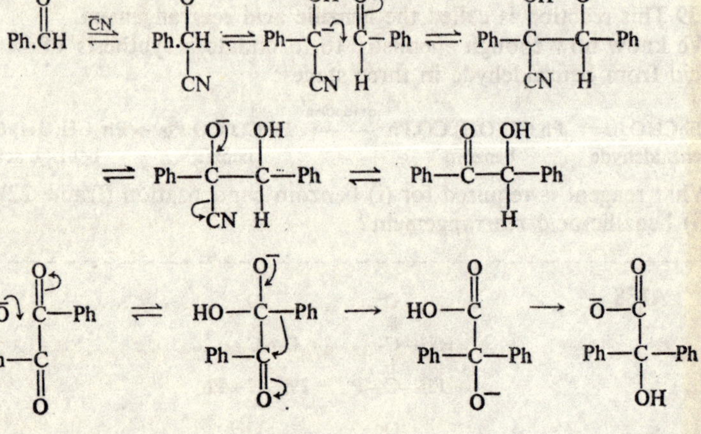

---

142 Test frames

Name the following derivatives of propanoic (propionic) acid and list them in decreasing order of reactivity:

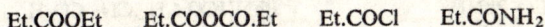

Et.COOEt   Et.COOCO.Et   Et.COCl $Et.CONH_2$

A141

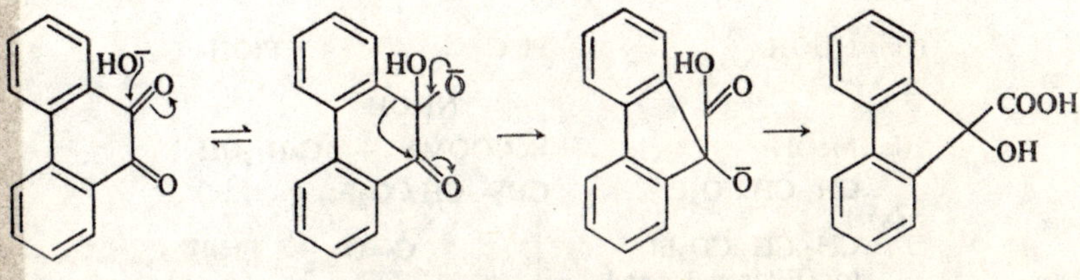

143 The following can be described as carbonyl substitution reactions. In each case give the structure of the attacking nucleophilic species and of the final products. The full reaction sequence is not required but write this out if it helps you to obtain the solution.

(i) ammonia and ethyl hexanoate,
(ii) acetyl chloride and sodium acetate,
(iii) benzoyl chloride, phenol, and sodium hydroxide,
(iv) succinic anhydride and aniline,
(v) methyl benzoate and sodium hydroxide,
(vi) propionic acid, ethanol, and hydrogen chloride,
(vii) methyl propionate and sodium,
(viii) ethyl benzoate, hydroxylamine, and potassium hydroxide,
(ix) hexyl propionate, methanol, and hydrogen chloride,
(x) ethyl adipate and sodium,
(xi) butyl acetate, methanol, and sodium.

A142

Et.COCl (propanoyl or propionyl chloride) > Et.COOCO.Et (propanoic or propionic anhydride) > Et.COOEt (ethyl propanoate or propionate) > $Et.CONH_2$ (propanamide or propionamide).

144 Give the product and the mechanism of the reaction in which sodium hydroxide reacts with the following aldehydes and ketones:

(i) $CH_3.CH_2.CHO$ (ii) *p*-$MeO.C_6H_4.CHO$ (iii) $Ph.CO.CCl_3$ (iv) Ph.CO.CO.Ph

A143

(i)	NH_3	$C_5H_{11}.CONH_2 + EtOH$
(ii)	$CH_3.COO^-$	$CH_3.CO.OCO.CH_3 + NaCl$
(iii)	PhO^-	$PhOCO.Ph + NaCl + H_2O$
(iv)	$PhNH_2$	$PhNHCO.CH_2.CH_2.CO_2H$
(v)	HO^-	$Ph.COO^- + MeOH$
(vi)	$EtOH$	$Et.COOEt + H_2O$
(vii)	$CH_3.\overset{-}{C}H.CO_2Me$	$CH_3.CH_2.CO.CH(CH_3).CO_2Me + MeOH$

(viii) $\overline{N}HOH$

$Ph.\overset{\overset{\displaystyle O}{\|}}{C} \ + \ EtOH$

$\qquad\qquad\qquad$ $NHOH$

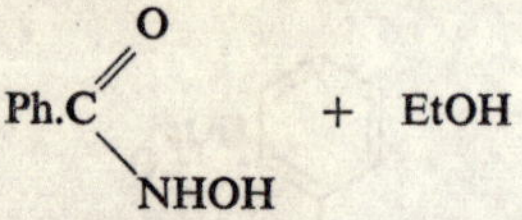

(ix)	$MeOH$	$Et.COOMe + C_6H_{13}OH$

(x) $\begin{array}{l} CH_2.\overset{-}{C}H.CO_2Et \\ | \\ CH_2.CH_2.CO_2Et \end{array}$ (or $\overset{-}{O}Et$ from Na and traces of EtOH)

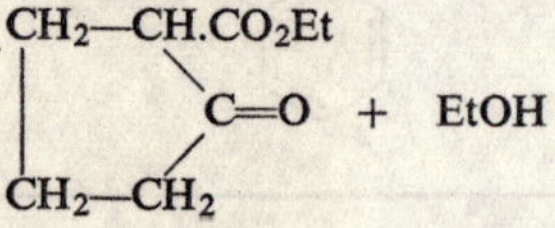

$\begin{array}{l} CH_2\!-\!CH.CO_2Et \\ | \qquad\quad C\!=\!O \ + \ EtOH \\ CH_2\!-\!CH_2 \end{array}$

(xi)	$Me\overset{-}{O}$	$CH_3.COOMe + C_4H_9OH$

A144

(i) $CH_3.CH_2.CHO \overset{\overline{O}H}{\rightleftharpoons} CH_3.\overset{-}{C}H.CHO \overset{Et.CHO}{\longrightarrow} Et.\overset{\overset{\displaystyle \overline{O} \quad CH_3}{}}{\underset{H}{C}}\!-\!CH.CHO$

$\overset{H_2O}{\longrightarrow} Et.CH(OH).CH(CH_3).CHO \overset{-H_2O}{\longrightarrow} Et.CH\!=\!C(CH_3).CHO$ (The product will be hydroxy-aldehyde or unsaturated aldehyde depending on the reaction conditions.)

(ii) $p\text{-}MeO.C_6H_4.CHO \overset{\overline{O}H}{\rightleftharpoons}$ $\rightleftharpoons$

$p\text{-}MeO.C_6H_4\!-\!\overset{\overset{\displaystyle O}{\|}}{\underset{OH}{C}} \ + \ H\overset{\overset{\displaystyle O^-}{|}}{\underset{H}{C}}.C_6H_4.OMe\text{-}p \ \rightarrow \ p\text{-}MeO.C_6H_4COO^-$

$\qquad\qquad\qquad\qquad\qquad\qquad + \ p\text{-}MeO.C_6H_4.CH_2OH$

(iii)

$HO\overset{\curvearrowright}{\searrow}\overset{\overset{\displaystyle O}{\|}}{\underset{Ph}{C}}\!-\!CCl_3 \rightleftharpoons HO\!-\!\overset{\overset{\displaystyle O^-}{|}}{\underset{Ph}{C}}\curvearrowright CCl_3 \rightleftharpoons HO\!-\!\overset{\overset{\displaystyle O}{\|}}{\underset{Ph}{C}} + \overline{C}Cl_3 \longrightarrow Ph.COO^- + CHCl_3$

(iv)

$HO\overset{\curvearrowright}{\searrow}\overset{\overset{\displaystyle O}{\|}}{\underset{Ph\!-\!C}{C}}\!-\!Ph \rightleftharpoons HO\!-\!\overset{\overset{\displaystyle O^-}{|}}{\underset{Ph\!-\!C}{C}}\!-\!Ph \rightleftharpoons HO\!-\!\overset{\overset{\displaystyle O}{\|}}{\underset{Ph\!-\!C}{C}}\!-\!Ph \longrightarrow \overset{\overset{\displaystyle O}{\|}}{\underset{OH}{C}}$

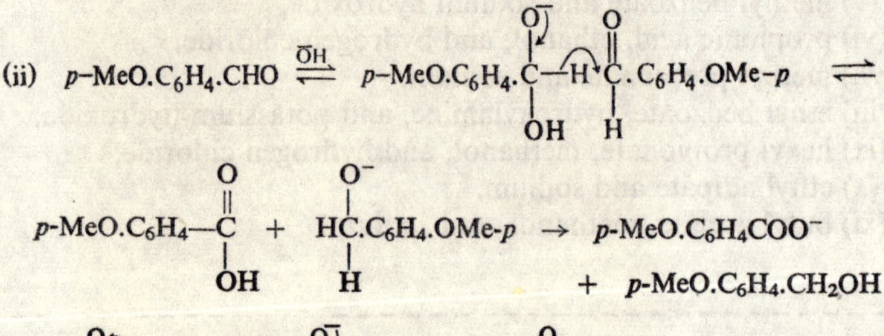

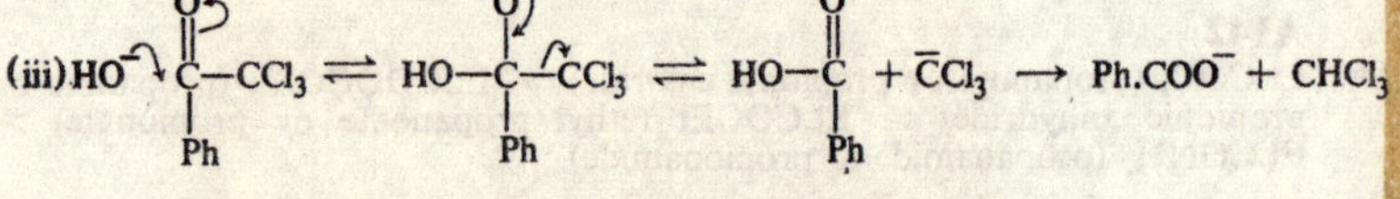

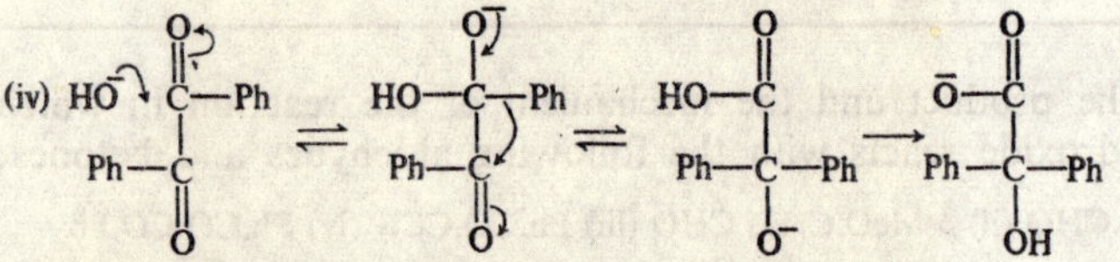